Brian Fleming Research & Learning Library
Ministry of Education
Ministry of Training, Colleges & Universities
900 Bay St. 13th Floor, Mowat Block
Toronto, ON M7A 1L2

D1787276

# Ethics for Citizenship in a Technological World

**Editor**
Roger B. Hill

**53rd Yearbook, 2004**
*Council on Technology Teacher Education*

New York, New York   Columbus, Ohio   Chicago, Illinois   Peoria, Illinois   Woodland Hills, California

 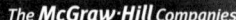

Copyright © 2004 by the Council on Technology Teacher Education.

All rights reserved. Except as permitted under the United States Copyright Act, no part of this publication may be reproduced or distributed in any form or by any means, or stored in a database or retrieval system, without prior written permission of the publisher, Glencoe/McGraw-Hill.

Send all inquiries to:
Glencoe/McGraw-Hill
3008 W. Willow Knolls Drive
Peoria, IL 61614

ISBN 0-07-865267-7

Printed in the United States of America.

1 2 3 4 5 6 7 8 9 10  044  08 07 06 05 04

Orders and requests for information about cost and availability of yearbooks should be directed to Glencoe/McGraw-Hill's Order Department, 1-800-334-7344.

Requests to quote portions of yearbooks should be addressed to the Secretary, Council on Technology Teacher Education, in care of Glencoe/McGraw-Hill at the above address, for forwarding to the current Secretary.

---

This publication is available in microform from

UMI
300 North Zeeb Road
Dept. P.R.
Ann Arbor, MI 48106

# FOREWORD

The CTTE Yearbook Series continues to represent an important forum for scholarly discourse on topics important to technology teacher education. Much of the success of the Council and profession is due to the kind of solid thinking and effort that are invested in the preparation of these yearbooks.

This year's yearbook focuses on a critically important topic for a technological world—ethics. As technology continues to expand and change, its interaction with social, political, and economic systems becomes increasingly complex. One of the more important aspects of these interactions has to do with ethical considerations, the ways of thinking, guidelines, and reflective capacity that should be used to guide and inform humane behavior in a civilized world. Interactions with technology are pervasive and complex. Daily, citizens and leaders are confronted with a wide range of ethical issues, such as the appropriate use and transfer of technology; the use of powerful communications technologies to facilitate, monitor, and inform; the role technology plays in the distribution of wealth on a global scale; the appropriate application of invasive medical and genetic technologies; and much more. Answers to these complex, technological issues are seldom simple and definitive. Constant changes in technology tend to present new and perplexing ethical issues. Further, globalization increasingly is bringing diverse and often conflicting sets of social, political, and cultural values and ideas to the discourse on technology.

The editor, Dr. Roger Hill, and the chapter authors are to be commended for their insightful treatment of this difficult topic. Although it is impossible to discuss the ethical dimensions of all technologies, the authors have done an excellent job of framing the key issues that should be addressed by technology educators. Those interested in technology often tend to focus on how technologies and technological systems work. Even with dramatic changes in technology in recent years, technology educators remain relatively comfortable with exploring topics such as "how things work" and how to improve the efficiency and functionality of technological systems. This is something that technology educators know how to do. This yearbook's thoughtful treatment of ethics extends the discourse into a more difficult and less well-defined area. However, the technological literacy of all citizens depends on the ability and willingness of technology educators to appropriately frame and intelligently discuss these important ethical issues.

Rodney L. Custer<br>
President, CTTE<br>
March 2004

# YEARBOOK PLANNING COMMITTEE

**Terms Expiring 2004**
   Cyril King
      South Eastern Education and Library Board
   John M. Ritz, Chairperson
      Old Dominion University
   Peter H. Wright
      Indiana State University

**Terms Expiring 2005**
   Patricia A. Hutchinson
      TIES Magazine
   Richard D. Seymour
      Ball State University

**Terms Expiring 2006**
   Patrick Foster
      Central Connecticut State University
   Edward M. Reeves
      Utah State University

**Terms Expiring 2007**
   Michael A. DeMiranda
      Colorado State University
   G. Eugene Martin
      Texas State University, San Marcos

**Terms Expiring 2008**
   Kurt R. Helgeson
      St. Cloud State University
   Linda Rae Markert
      State University of New York at Oswego

# OFFICERS OF THE COUNCIL

**President**
Rodney L. Custer
Illinois State University
Industrial Technology Department
Normal, IL 61790-5100

**Vice President**
Richard D. Seymour
Ball State University
Department of Industry and Technology
Muncie, IN 47306

**Secretary**
Michael K. Daugherty
Illinois State University
Department of Technology
Normal, IL 61790-5100

**Treasurer**
Marie C. Hoepfl
Appalachian State University
Department of Technology
Boone, NC 28608

**Past President**
John M. Ritz
Old Dominion University
Occupational and Technical Studies
Norfolk, VA 23529-0498

# YEARBOOK PROPOSALS

Each year at the ITEA International Conference, the CTTE Yearbook Committee reviews the progress of yearbooks in preparation and evaluates proposals for additional yearbooks. Any member is welcome to submit a yearbook proposal, which should be written in sufficient detail for the committee to be able to understand the proposed substance and format. Fifteen copies of the proposal should be sent to the committee chairperson by February 1 of the year in which the conference is held. The following criteria are employed by the committee in making yearbook selections.

CTTE Yearbook Committee

## CTTE Yearbook Guidelines

A. **Purpose**

The CTTE Yearbook Series is intended as a vehicle for communicating major topics or issues related to technology teacher education in a structured, formal series that does not duplicate commercial textbook publishing activities.

B. **Yearbook Topic Selection Criteria**

An appropriate yearbook topic should:
1. Make a direct contribution to the understanding and improvement of technology teacher education.
2. Add to the accumulated body of knowledge of technology teacher education and to the field of technology education.
3. Not duplicate publishing activities of other professional groups.
4. Provide a balanced view of the theme and not promote a single individual's or institution's philosophy or practices.
5. Actively seek to upgrade and modernize professional practice in technology teacher education.
6. Lend itself to team authorship as opposed to single authorship.

Proper yearbook themes related to technology teacher education may also be structured to:
1. Discuss and critique points of view that have gained a degree of acceptance by the profession.
2. Raise controversial questions in an effort to obtain a national hearing.
3. Consider and evaluate a variety of seemingly conflicting trends and statements emanating from several sources.

C. **The Yearbook Proposal**
1. The yearbook proposal should provide adequate detail for the Yearbook Committee to evaluate its merits.
2. The yearbook proposal includes the following elements:
   a) Define and describe the topic of the yearbook.
   b) Identify the theme and describe the rationale for the theme.
   c) Identify the need for the yearbook and the potential audience or audiences.
   d) Explain how the yearbook will advance the technology teacher education profession and technology education in general.
   e) Diagram symbolically the intent of the yearbook.
   f) Provide an outline of the yearbook that includes:
      i) A table of contents
      ii) A brief description of the content or purpose of each chapter
      iii) At least a three-level outline for each chapter
      iv) Identification of chapter author(s) and backup authors
      v) An estimated number of pages for each yearbook chapter
      vi) An estimated number of pages for the yearbook (not to exceed 250 pages)
   g) Provide a timeline for completing the yearbook.

It is understood that each author of a yearbook chapter will sign a CTTE Editor/Author Agreement and comply with the Agreement. Additional information on yearbook proposals is found on the CTTE Web site at http://teched.vt.edu/ctte/.

# PREVIOUSLY PUBLISHED YEARBOOKS

*1. *Inventory Analysis of Industrial Arts Teacher Education Facilities, Personnel and Programs,* 1952.
*2. *Who's Who in Industrial Arts Teacher Education,* 1953.
*3. *Some Components of Current Leadership: Techniques of Selection and Guidance of Graduate Students; An Analysis of Textbook Emphases,* 1954, three studies.
*4. *Superior Practices in Industrial Arts Teacher Education,* 1955.
*5. *Problems and Issues in Industrial Arts Teacher Education,* 1956.
*6. *A Sourcebook of Reading in Education for Use in Industrial Arts and Industrial Arts Teacher Education,* 1957.
*7. *The Accreditation of Industrial Arts Teacher Education,* 1958.
*8. *Planning Industrial Arts Facilities,* 1959. Ralph K. Nair, ed.
*9. *Research in Industrial Arts Education,* 1960. Raymond Van Tassel, ed.
*10. *Graduate Study in Industrial Arts,* 1961. R. P. Norman and R. C. Bohn, eds.
*11. *Essentials of Preservice Preparation,* 1962. Donald G. Lux, ed.
*12. *Action and Thought in Industrial Arts Education,* 1963. E. A. T. Svendsen, ed.
*13. *Classroom Research in Industrial Arts,* 1964. Charles B. Porter, ed.
*14. *Approaches and Procedures in Industrial Arts,* 1965. G. S. Wall, ed.
*15. *Status of Research in Industrial Arts,* 1966. John D. Rowlett, ed.
*16. *Evaluation Guidelines for Contemporary Industrial Arts Programs,* 1967. Lloyd P. Nelson and William T. Sargent, eds.
*17. *A Historical Perspective of Industry,* 1968. Joseph F. Luetkemeyer Jr., ed.
*18. *Industrial Technology Education,* 1969. C. Thomas Dean and N. A. Hauer, eds.; *Who's Who in Industrial Arts Teacher Education,* 1969. John M. Pollock and Charles A. Bunten, eds.
*19. *Industrial Arts for Disadvantaged Youth,* 1970. Ralph O. Gallington, ed.
*20. *Components of Teacher Education,* 1971. W. E. Ray and J. Streichler, eds.
*21. *Industrial Arts for the Early Adolescent,* 1972. Daniel J. Householder, ed.
*22. *Industrial Arts in Senior High Schools,* 1973. Rutherford E. Lockette, ed.
*23. *Industrial Arts for the Elementary School,* 1974. Robert G. Thrower and Robert D. Weber, eds.
*24. *A Guide to the Planning of Industrial Arts Facilities,* 1975. D. E. Moon, ed.
*25. *Future Alternatives for Industrial Arts,* 1976. Lee H. Smalley, ed.
*26. *Competency-Based Industrial Arts Teacher Education,* 1977. Jack C. Brueckman and Stanley E. Brooks, eds.
*27. *Industrial Arts in the Open Access Curriculum,* 1978. L. D. Anderson, ed.
*28. *Industrial Arts Education: Retrospect, Prospect,* 1979. G. Eugene Martin, ed.
*29. *Technology and Society: Interfaces with Industrial Arts,* 1980. Herbert A. Anderson and M. James Benson, eds.
*30. *An Interpretive History of Industrial Arts,* 1981. Richard Barella and Thomas Wright, eds.

*31. *The Contributions of Industrial Arts to Selected Areas of Education,* 1982. Donald Maley and Kendall N. Starkweather, eds.
*32. *The Dynamics of Creative Leadership for Industrial Arts Education,* 1983. Robert E. Wenig and John I. Mathews, eds.
*33. *Affective Learning in Industrial Arts,* 1984. Gerald L. Jennings, ed.
*34. *Perceptual and Psychomotor Learning in Industrial Arts Education,* 1985. John M. Shemick, ed.
*35. *Implementing Technology Education,* 1986. Ronald E. Jones and John R. Wright, eds.
 36. *Conducting Technical Research,* 1987. Everett N. Israel and R. Thomas Wright, eds.
*37. *Instructional Strategies for Technology Education,* 1988. William H. Kemp and Anthony E. Schwaller, eds.
*38. *Technology Student Organizations,* 1989. M. Roger Betts and Arvid W. Van Dyke, eds.
*39. *Communication in Technology Education,* 1990. Jane A. Liedtke, ed.
*40. *Technological Literacy,* 1991. Michael J. Dyrenfurth and Michael R. Kozak, eds.
 41. *Transportation in Technology Education,* 1992. John R. Wright and Stanley Komacek, eds.
*42. *Manufacturing in Technology Education,* 1993. Richard D. Seymour and Ray L. Shackelford, eds.
*43. *Construction in Technology Education,* 1994. Jack W. Wescott and Richard M. Henak, eds.
 44. *Foundations of Technology Education,* 1995. G. Eugene Martin, ed.
*45. *Technology and the Quality of Life,* 1996. Rodney L. Custer and A. Emerson Wiens, eds.
 46. *Elementary School Technology Education,* 1997. James J. Kirkwood and Patrick N. Foster, eds.
 47. *Diversity in Technology Education,* 1998. Betty L. Rider, ed.
 48. *Advancing Professionalism in Technology Education,* 1999. Anthony F. Gilberti and David L. Rouch, eds.
*49. *Technology Education for the 21st Century: A Collection of Essays,* 2000. G. Eugene Martin, ed.
 50. *Appropriate Technology for Sustainable Living,* 2001. Robert C. Wicklein, ed.
 51. *Standards for Technological Literacy: The Role of Teacher Education,* 2002. John M. Ritz, William E. Dugger, and Everett N. Israel, eds.
 52. *Selecting Instructional Strategies for Technology Education,* 2003. Kurt R. Helgeson and Anthony E. Schwaller, eds.

\* Out-of-print yearbooks can be obtained in microfilm and in Xerox copies. For information on price and delivery, write to UMI, 300 North Zeeb Road, Dept. P.R., Ann Arbor, Michigan 48106.

# PREFACE

The publication of this yearbook represents not only the culmination of hundreds of hours of diligent work, but also is a milestone in the ongoing research stream of the editor. When the suggestion was made that I consider submitting a proposal for a CTTE yearbook, it was only natural to turn to the theme that I have continued to focus on over the years. That theme could be characterized as focusing on the interrelationship between technological competence and ethical behavior within a world permeated by technology.

Much of my own prior work related to work ethic and affective attributes that contribute to career success. Dependability, initiative, and interpersonal skills have emerged as key characteristics of people who function well within our technological world. Work ethic overlaps with the larger context of ethics, however, and it is this broader scope that provided a basis for this book. In many ways, this book is about making choices that allow us and others to be successful in life.

Preparation of this volume has been an educational process. I have come to appreciate the complexity of the topic and the diversity of approaches in dealing with it. All of the authors who contributed to this work strongly believe in the importance of ethics, and we have labored to represent a wide range of perspectives on the issues covered. Geographically we are scattered across the United States, and our backgrounds are culturally diverse. In each case, however, life experiences have taught us the importance of ethics and the huge risks involved when technology is not ethically managed.

Regardless of where a person lives, it would be difficult to have escaped the prominence of ethics as an important issue over the past several years. Week after week, the news media provides examples of persons who have either behaved unethically or who have been accused of such. Problems in the area of ethical behavior have been widely publicized in the corporate world, in the world of sports, in the world of politics, and even in the world of religion. Why all the attention? Why is ethics such an important topic?

The reason ethics is such an important issue is that our technology has empowered individual people like never before in the history of the world. Information and communications technologies have especially contributed to this empowerment. Consider the Internet, for example, and an individual end user connected to that vast resource. This person has the power to find all sorts of useful information that could help with a school project, locate medical advice related to the illness of a loved one, or conduct electronic transactions to make purchases or manage finances. This same person also has the power to transmit computer viruses to other Internet users, to disseminate false or harmful information, and to conduct dishonest business transactions that could impact numerous other people. In principle, people have always had the power to do good or bad, but technology now greatly magnifies the potential impact of these ethical decisions.

Publishing a book about ethics, and particularly suggesting that ethics should be a part of the content to be taught in public schools, is not without risk of criticism. An underlying premise of this book is that integrity, responsibility, fairness, caring, initiative, interpersonal skills, and dependability are attributes that all people accept as desirable. Although that might be the case, this book also assumes it is appropriate for schools to encourage the development of these attributes in students.

On this latter issue, some might argue that the teaching of character traits, personal behavior, and values belongs in the home and is the responsibility of parents. In some parts of the country, school systems have been criticized for including character education in their curricular programs or for addressing morals and values as a part of school instruction. The truth of the matter is, most states include global curricular objectives stating that schools *will* develop good character and encourage high moral standards in all instructional programs. It is the details of implementation that sometimes creates controversy.

Another source of potential criticism of school instruction that encourages integrity, responsibility, fairness, caring, initiative, interpersonal skills, and dependability is that this might be done for the benefit of future employers rather than for the personal benefit of students. The truth of the matter is that employers are among the most vocal proponents of having students develop these attributes. They have a lot of difficulty finding employees who arrive at work on time, get along with fellow employees, and manage time wisely while on the job. Success at work, however, can also be beneficial to the employee. Whether through advancement, satisfaction with a job well done, or avoiding the ire of management, employees usually have a more enjoyable work experience when they practice ethical behavior.

The primary or target audience for this book is composed of undergraduate and graduate technology education majors, technology teacher educators, and technology teachers. The *Standards for Technological Literacy: Content for the Study of Technology,* released in March of 2000 by the International Technology Education Association with funding from the NSF and NASA, acknowledge that technologically literate citizens must have knowledge that extends beyond the design and operation of technological systems. Cultural, social, economic, and political effects of technology and ethical issues in the development and use of technology are specifically listed as content that should be covered by middle grades and high school technology education programs along with other school coursework contributing to technologically literate students.

Implementing the new technology education standards requires some adjustments in typical instructional activities. New materials and resources are required. This is especially true with regard to some of the content related to the study of technology and society. A wide array of materials and instructional strategies already exist for much of the other content identified in the standards—understanding the nature of technology, knowing about and being able to use the design process, and understanding a variety of technological systems. New resources are needed for teachers to use in addressing topics related to technology and society, including the relevant ethical issues identified within the standards.

One of the concerns teachers raise when standards specify additional content for their subject areas is where the time will come from to teach these new topics. It is often necessary to combine topics within a single lesson. For example, using a case study that combines elements of designing a transportation system with an ethical problem-solving activity can accomplish multiple objectives. Learning is often enhanced as content is situated and taught using true-to-life scenarios.

It was the desire of the authors who collectively developed the materials in this book that educators involved in developing technological literacy be equipped with a resource to help them include ethics as a part of their instructional programs. With that end in mind, a variety of kinds of materials are included in the following pages.

In Chapter 1, a definition of ethics is provided along with a discussion of basic human development and sociological theories relevant to ethical belief systems and attitudes. Motives for ethical behavior are discussed ranging from personal commitments to financial incentives. This chapter sets the stage for understanding ethics and ethical issues relevant to technological literacy. It also provides a rationale for the importance of ethics as a component of technological literacy.

Materials in Chapter 2 deal with the impact Euro-American thought has had on the United States and other developed nations. Perspectives of other cultures are explored along with the ways in which associated ethics have impacted technological development. Content in this chapter has particular relevance to Standard 6, the role of society in the development and use of technology, and Standard 7, the influence of technology on history, from the *Standards for Technological Literacy: Content for the Study of Technology.*

The content of Chapter 3 includes a series of case studies that highlight the role of ethics and character in the design and development of technological systems. The objective here was to facilitate understanding of the significance of ethics in a technological workplace and to highlight the dilemmas and problems that can occur when ethics are absent.

The focus of Chapter 4 is the role of ethics in assessing technological impacts on society and the importance of ethics in development of sustainable forms of technology. When work with technological systems is approached in an atmosphere devoid of moral and ethical considerations, the outcomes are often detrimental to the long-term well-being of mankind. Consideration was given to Standard 4, the cultural, social, economic, and political effects of technology; Standard 5, the effects of technology on the environment; and Standard 6, the role of society in the development and use of technology, from the *Standards for Technological Literacy: Content for the Study of Technology* as this chapter was developed.

Chapter 5 provides a conceptual framework for human development, and describes the importance of adopting a well-thought-out belief system related to moral and ethical issues. Practical strategies are presented for incorporating discussion of these topics within technology education instruction, without crossing the bounds of legal and political constraint. This chapter also addresses Standards 4, 5, 6, and 7 from the *Standards for Technological Literacy: Content for the Study of Technology*.

Chapter 6 provides a review for existing instructional materials commonly used in technology education programs that address ethics and related issues. This chapter also provides some discussion of the treatment of ethics in technology education literature as well as professional ethics for technology education professionals at all levels of the educational system.

Content included in Chapter 7 provides examples of learning activities that could be integrated into instruction related to medical technologies, agricultural and related biotechnologies, energy and power technologies, information and communication technologies, transportation technologies, manufacturing technologies, and construction technologies. Design and problem solving are an integral part of the activities described, and the problems presented have embedded ethical dilemmas. This chapter addresses multiple items from the *Standards for Technological Literacy: Content for the Study of Technology*, including Standard 11, apply the design process, and Standard 13, assess the impact of products and systems.

Chapter 8 summarizes the key elements pertinent to ethics as a component of technological literacy and addresses quality of workforce issues in economic development. One of the typical outcomes expected of any educational endeavor is to equip students to lead successful and fulfilling lives. This not only involves an ability to make good personal decisions, but also to be good citizens in communities and workplaces.

Chapter 9 provides a summary for the book and provides some final suggestions for how these materials might be used by the profession. Some recommendations are also made for future work related to ethics and technological literacy.

Although numerous other topics and an endless array of related materials could have been added to the contents of this book, the material that is provided should provide a good start for those seeking guidance to include ethics as a component of technology instruction. An additional resource that has been developed is a companion Web site. In several instances, the authors were aware of online materials that were related to the chapter contents. Placing online locations in print, however, always runs a risk of including resources that are no longer available. By placing these links on a Web site, they can be maintained, added to, and enhanced over time. The address for the companion Web site is http://www.uga.edu/teched/ethics.

# ACKNOWLEDGMENTS

The work presented in this book was only possible because the editor and authors stand on the shoulders of so many dedicated educators who came before—educators with a strong commitment to ethics and belief in the importance of teaching. Our values and ethical belief systems were influenced, beginning at an early age, by parents, friends, and adults with whom we came into contact. This book is dedicated to all who came before us who were steadfast in their convictions about the importance of ethical decision making.

The editor and authors, in addition to the members and officers of the Council on Technology Teacher Education, would like to acknowledge Glencoe/McGraw-Hill for its continued support of the Council's Yearbook Series. This yearbook, *Ethics for Citizenship in a Technological World,* is the Council's 53rd Yearbook Edition. Without the support of Glencoe/McGraw-Hill and the efforts of Wes Coulter, Trudy Muller, and Jean Leslie, educators endeavoring to encourage technological literacy would lose a valuable source of instructional materials and an important resource for professional development.

The editor would like to thank Gene Martin, who provided encouragement and mentoring during the early stages of this project, and other members of the CTTE Yearbook Committee, who decided that this topic was a significant one for the profession. The editor also acknowledges the contributions of each author and expresses thanks for their commitment of time and energy to prepare the chapters included in this book.

A final word of thanks goes to Joan Taylor at the University of Georgia, who read every word of this book and provided a thorough review and editing for this work.

                                              53rd Yearbook Editor
                                              Roger B. Hill

# TABLE OF CONTENTS

Foreword ................................................................................................................. iii
Yearbook Planning Committee .......................................................................... iv
Officers of the Council ......................................................................................... v
Yearbook Proposals ............................................................................................. vi
Previously Published Yearbooks ...................................................................... vii
Preface .................................................................................................................. ix
Acknowledgments ............................................................................................ xiv

**Chapter 1: Introduction to Ethical Issues in a Technological World ............... 1**
  Roger B. Hill
  The University of Georgia
  Athens, GA

**Chapter 2: Ethics in a Culturally Diverse Technological World ..................... 21**
  Linda Rae Markert
  State University of New York at Oswego
  Oswego, NY

**Chapter 3: Ethics and the Design and Development of
       Technological Systems ..................................................................... 49**
  Michael A. DeMiranda
  Colorado State University
  Fort Collins, CO

  Len S. Litowitz
  Millersville University
  Millersville, PA

  Mark Sanders
  Virginia Tech
  Blacksburg, VA

  Richard D. Seymour
  Ball State University
  Muncie, IN

  Jack Wescott
  Ball State University
  Muncie, IN

*xv*

*Myra N. Womble and Stephanie Williams*
*The University of Georgia*
*Athens, GA*

**Chapter 4: Ethics and the Assessment of Technological Impacts on Society** ......................................................................... 123

*Robert C. Wicklein*
*The University of Georgia*
*Athens, GA*

**Chapter 5: Developmental and Contextual Issues Related to Ethics and Character** ................................................................... 145

*Rodney L. Custer*
*Illinois State University*
*Normal, IL*

*Danny C. Brown*
*Illinois State University*
*Normal, IL*

**Chapter 6: The Status of Ethics in Technology Education** ........................... 163

*Philip A. Reed*
*Old Dominion University*
*Norfolk, VA*

*Susan Presley*
*North Cobb High School*
*Kennesaw, GA*

*Angela Hughes*
*Morrow High School*
*Morrow, GA*

*Diane Irwin Stephens*
*Jasper County Middle/High Schools*
*Monticello, GA*

**Chapter 7: Ethics and the Study of the Designed World ...............187**
  Michael A. DeMiranda and Nick Benson
  Colorado State University
  Fort Collins, CO

  Len S. Litowitz
  Millersville University
  Millersville, PA

  Mark Sanders
  Virginia Tech
  Blacksburg, VA

  Richard D. Seymour
  Ball State University
  Muncie, IN

  Jack Wescott
  Ball State University
  Muncie, IN

  Myra N. Womble and Stephanie Williams
  The University of Georgia
  Athens, GA

**Chapter 8: Ethics in a Global Economic System ...........................243**
  Archie B. Carroll
  The University of Georgia
  Athens, GA

**Chapter 9: Closing Thoughts about Ethics for Citizenship in a Technological World ..................................................................267**
  Roger B. Hill
  The University of Georgia
  Athens, GA

# Introduction to Ethical Issues in a Technological World

## Chapter 1

Roger B. Hill
The University of Georgia
Athens, GA

Clichés about technology and the twenty-first century abound. Many of them refer to the marvels of modern technology and the ways in which it empowers people to live out their lives. Whether in areas related to discovering new cures for disease or developing more efficient systems of transportation, advances in technology impact people and shape the context in which they exist. Choices about when and how to use technology largely determine whether technological impacts are helpful or harmful, and these decisions often have ethical or value-laden components.

This book is about ethics, values, and citizenship in a technological world. It was developed based on the premise that all persons should be technologically literate and equipped to deal with ethical issues that are part of problem solving and design decisions. This first chapter provides a definition of ethics along with a discussion of basic human development and sociological theories relevant to ethical belief systems and values. A model for making ethical decisions is presented and explained. This chapter sets the stage for understanding ethics and ethical issues relevant to technological literacy. It also provides a rationale for the importance of ethics as a component of technological literacy.

## WHAT ARE ETHICS AND VALUES?

The dictionary defines ethics as (*a*) "a set of moral principles or values," (*b*) "a theory or system of moral values," (*c*) "the principles of conduct governing an individual or a group," or (*d*) "a guiding philosophy" (Merriam-Webster, 1998). Values are defined as "something (as a principle or quality) intrinsically valuable or desirable." In practice, ethics guide the process of choosing a right course of action within the context of interactions with other people and institutions. Values reflect individual beliefs and desires and are illuminated by the ethical choices people make.

Ethical principles and values are developed by individuals based on influences at home, in school, and through participation in community or religious organizations. Life experiences also influence ethics. Television and other media impact the development of ethics and shape perceptions of acceptable practice. Authority, culture, intuition, and reason can each serve as a basis for ethical decisions (Goree, 1996). These sources for ethical principles can work in concert, but some of these are typically more influential than the others in an individual's ethical decision-making processes.

When an action is determined to be right or wrong because someone said so, authority is the basis for the ethical stance. Authority is at work when something is wrong because "God forbids it" as well as when a government leader or institution declares something wrong and ethical behavior follows. Obeying speed limit laws because it is against the law to drive faster than the speed limit is an example of ethics derived from authority.

Ethical beliefs can also be attributed to a person's culture. Even though there are some ethical standards that are pervasive in all cultures, others can vary from one people group to another. If a person believes it is right to drive as much as five miles per hour faster than the speed limit because it is an accepted practice among the other people with whom the person has come into contact, this is an example of an ethical decision based on culture.

Intuition or conscience is another source of influence on ethical behavior. The role varies from person to person, and some part of this factor is based on integration of values into a person's subliminal belief systems. Guilt also plays a role because decisions are made to avoid this feeling. If a person avoids exceeding a speed limit because of guilty feelings or intuitively realizing speeding is wrong, this is an example of conscience as a source of ethical decision making.

Reason can serve as a basis for ethical decisions. If a person weighs the arguments on both sides of an ethical issue and chooses to act based on rationale rather than on authority, culture, or conscience, reason is the source of the ethical decision. A person who drives at a speed equal to or less than the posted speed limit based on a rationale that accidents are more likely to occur at higher speeds is making an ethical choice based on reason.

## PHILOSOPHY OF ETHICS

The field of philosophy provides several theoretical frameworks for making ethical decisions. These theories are well-developed and encompass distinct perspectives. Consideration of ethics and values would be incomplete without discussion of ethical relativism, utilitarianism, and deontological, justice, and virtue theories. The first of these is a prevalent but flawed perspective. The others provide viable frames for ethical positions, arguments, and critical examination, a process ethicists refer to as dialectic.

Ethical relativism asserts that there are no universal moral norms and that what is right for one person is not necessarily right for another. Arguments for the position often point to acceptable practices that vary from one society to another. Examples of societies can be identified in which polygamy, slavery, or infanticide are accepted, whereas in other societies, these acts are morally wrong. In some societies, women cover their heads and faces in public, whereas in other societies, they do not. Acceptable practice for interactions between men and women varies considerably among different cultures. Examples of these types are used in arguments supporting ethical relativism.

Using arguments that focus on specific practices to support an ethical position is problematic. In the fifteenth century, some people believed that the sun orbited around the earth, and others believed the earth moved around the sun. This is a descriptive statement, a statement of fact that can be investigated and proven to be correct or incorrect. Ethical relativists make a normative claim based on the diversity of moral standards in different societies. Normative statements prescribe what people should do to be ethical. Making a normative claim that there are no universal moral norms based on diverse societal examples is like stating that the orbital paths in the solar system are relative if different people believe different things.

Utilitarianism is an ethical framework that determines whether action is right or wrong based on outcomes or consequences. Moral decisions should be based on producing the greatest amount of good for the greatest number of people. Johnson (2001) posits that good is often equated with happiness and that happiness is the ultimate intrinsic good. The goal of utilitarianism is not maximizing individual happiness, but the total happiness of all persons affected by a decision. The happiness of a few might be subrogated to the happiness of a larger group.

The basis for legislation and policy decisions in contemporary democratic societies is generally considered to be utilitarianism. Government action sometimes results in unhappiness or detrimental consequences for an individual or group of persons, but the outcomes are favorable for the majority.

Utilitarianism can be criticized on the basis of imposing unfair burdens on a few for the sake of many. Imposing slavery on a few persons in society might result in great happiness for many, but humans have an internal moral intuition that indicates slavery is wrong. Utilitarian arguments might be made in support of atrocities related to medical research committed in World War II Germany, but moral conscience again indicates that harming a few for the supposed good of many is not right. Utilitarian philosophers counter that these examples represent short-sighted approaches and that long-term consequences, guiding utilitarian moral decision making, would never justify atrocities.

Another ethical framework for making moral decisions is that of the deontologist. This perspective is based on the duty of people to respect the value of other human beings. Humans are distinct from other creatures because of their ability to make rational decisions. Unlike other living organisms, humans can legislate or create rules to be followed that benefit others when everyone obeys them.

Deontologists establish rules for making moral decisions by considering whether the greatest good would come if everyone in the world behaved in that manner. Duty to act in principled ways might lead to the same behaviors chosen by universalists in some circumstances, but the reasoning would be different. Whereas the universalist makes choices based on the most good for the most people, the deontologist makes choices based on principles rather than trying to determine potential consequences.

Justice and the idea of people having rights provide another facet for consideration as dialectic takes place. Rights are sometimes associated with deontological perspectives, but arguments can be made for rights based on other ethical perspectives as well. An argument for a right to protection of one's intellectual property could be made from a utilitarian perspective.

Several distinctions can be made when considering rights. Johnson (2001) distinguishes between negative rights and positive rights. Negative rights restrain actions by others that would cause harm. The right to be allowed to live and not be intentionally harmed by others is an example of a negative right. A positive right requires action by people to sustain the life and well-being of others. The principle of positive rights can contribute to charitable activities and efforts to aid others who have met with misfortune.

Another distinction in rights can be made between legal rights and human rights. Legal rights are established by law, but human rights are associated with the uniqueness of human existence. Humans have the capacity to act in ways that would be harmful to others or to cooperate so that the health and safety of all members of a group are protected. When humans agree, either implicitly or explicitly, to cooperate for the betterment of the group, a social contract is formed. Human rights are derived from social contracts and obligate people to submit to rules of civility. The social contract creates expectations about how people will treat one another, and these expectations comprise a person's rights.

Virtue ethics embody a frame of thinking that focuses on desirable, personal qualities. Virtue, coming from the Greek word areté meaning quality or excellence, is more related to moral character than to a system for making ethical decisions. The result of virtue, however, is ethical, care-based behavior. People characterized by virtue are benevolent, generous, tolerant, and unselfish. Discussion of virtue and related character traits can be traced to the early Greek philosophers, but virtue ethics and care-based thinking are often associated with Christianity. The Golden Rule, treating others as one would want to be treated, is a central theme of the behavior that results from virtue ethics.

Some philosophers have argued that virtue and care-based thinking do not provide the practical principles necessary for an ethical framework (Kidder, 1996), but most of the world's religious belief systems embrace virtue as a central tenet. The influence of religious beliefs has been a significant factor in the history of mankind, and virtue ethics should not be marginalized. For some people, a relationship with God is central to everything they do. Virtue ethics and care-based thinking provide the best framework for examining or explaining these people's ethical decision-making processes.

## MORAL DEVELOPMENT THEORY

Ethical perspectives can vary considerably from person to person. Someone who approaches ethical decisions using utilitarianism often views things quite differently than the person characterized by deontological thinking. Diversity in ethical frames of reference can actually be an asset in a democratic society because the expression of differing perspectives encourages examination of all aspects of a decision.

In addition to differing philosophies, people can differ developmentally. Many factors can influence a person's capacity to cope with ethical issues, but the work of Kohlberg (1975) can be used to describe how people develop in this area. He has described six stages of moral development grouped into three levels. Active thinking about moral issues and decisions stimulates movement through these moral stages. Kohlberg also noted that people at different levels of development respond to ethical issues in different ways.

Kohlberg's *preconventional level* describes people who can identify "right" and "wrong" but interpret these labels based on external forces. Decisions involving ethical issues are based on avoiding punishment or obtaining some type of reward from those in positions of power. Fairness and reciprocity are recognized on a superficial level, but this *quid pro quo* is not based on loyalty or gratitude.

Kohlberg's *conventional level* represents stages of moral development involving loyalty to family and support for social order. Ethical decisions are based on pleasing or helping others, and conformity to majority behavior is valued. This level also encompasses a law-and-order orientation, in which authority is respected and the importance of maintaining social order is recognized.

The *postconventional level* includes Kohlberg's most sophisticated stages of moral development. For people at this level, choices involving ethics take into account the perspectives and needs of all persons involved. Individual rights and standards that have been critically examined and approved by the society through democratic processes influence right action. The relativism of personal values and opinions is recognized, but emphasis is placed on procedural rules and consensus. This level of moral development also embraces universal principles of justice, reciprocity and equality of human rights, and the value and dignity of all people.

Gilligan (1982) identified several weaknesses in the application of Kohlberg's model to the moral development of women. Sometimes referred to as an ethic of care, as compared to Kohlberg's ethic of justice, Gilligan described the moral development of women as being heavily influenced by the value they tend to place on relationships. Women tend to focus on developing intimacy and relationships, but men focus more on autonomy, competition, and fairness.

## CAN ETHICS BE TAUGHT?

Kohlberg (1975) attributed the first fully developed theoretical basis for moral education to John Dewey. Jean Piaget had earlier defined stages of moral development in children based on studies of cognitive developmental stages. Dewey stated that the goal of education was intellectual and moral development, and he postulated three levels of moral development similar to those described by Kohlberg. Ethical and psychological principles provided a basis for character development, according to Dewey, and an understanding of these was viewed as essential to success in developing character in students.

Moral development and acquisition of ethical decision-making skills begin very early in life. Early growth in this area is largely dependent on family and community influences during infancy and childhood. Young people detect whether a coherent ethical philosophy is being applied by parents and other adults by observing the level of consistency. When children are able to observe ethical decision making that is consistent from one situation to another, they tend to be more secure in their environment.

Schools have been engaged to provide opportunities for character development as well as acquisition of knowledge and skills. As institutions such as the nuclear family unit have become less permanent, society has transferred greater and greater expectations on schools (Vincent and Meche, 2001). Meeting the challenge to teach ethics has been difficult for many educators because of lack of formal preparation and the potential for criticism from agents within local communities who are concerned about the source of ethical principles that might be taught (Tucker and Stout, 1991).

A question that is sometimes raised is whether ethics can be taught to adolescents or adults. If ethical behavior was not learned during childhood and moral development is at a preconventional level, it can be daunting to teach ethics to adolescent or adult learners. Effective strategies typically include group interaction with engaging real-life scenarios that stimulate discussion and reflection. Learners must choose for themselves whether to adopt new practices or change ethical decision-making processes, but awareness of consequences can definitely be conveyed.

The content to be taught is another issue related to ethics instruction. Sources of influence can be identified, and ethical philosophies can be discussed. Educational objectives might also include an expectation that learners be encouraged to adopt standards of behavior associated with virtue. Public school professionals must use care to establish secular arguments for the constructs they choose to emphasize because virtue ethics are often associated with religious belief systems. The framework for topics being taught should be grounded in scholarly research so that a clear rationale can be provided in response to critics who raise questions about public school ethics instruction.

## UNIVERSAL VALUES

Definitions of ethics and higher levels of moral development include references to moral principles. One of the issues that is relevant to any discussion of ethics is whether there are universal ethical principles that all people accept. The values clarification movement of the 1970s encouraged people to discover their own values without providing an endorsement of any universal principles, but it failed to provide an effective strategy for encouraging ethical behavior. The values clarification movement was built on the premise that it is not appropriate for those in positions of power to indoctrinate others to adopt currently fashionable values (Kinnier, Dautheribes, and Therese, 2000). Ethical relativism was a popular philosophy in the United States during the 1970s, and the values clarification movement was shaped by this perspective.

As the twentieth century ended and the twenty-first century began, the decline of ethical behavior in the workplace, entertainment, politics and government, and numerous other societal contexts resulted in several initiatives to identify a set of universal, ethical principles (Kidder, 1994; Kinnier, Dautheribes, and Therese, 2000; Nish, 1996). School reform recommendations such as the SCANS Report for America 2000 (Secretary's Commission on Achieving Necessary Skills, 1992) included personal qualities (individual responsibility, self-esteem, sociability, self-management, and integrity) in the list of outcomes that educational programs should seek to achieve.

Kidder (1994) developed a list of eight common values that transcended international borders and cultural traditions. Based on interviews with 24 individuals in 16 nations, he identified love, truthfulness, fairness, freedom, unity, tolerance, responsibility, and respect for life as values people in all cultures espouse.

Another initiative to identify a list of shared values was undertaken by the Josephson Institute of Ethics. In 1992, they assembled a diverse group of education and youth service leaders in Aspen, Colorado (Nish, 1996). The task of this group was to find consensus on values that all Americans could agree on regardless of political persuasion, religious views, race, ethnicity, or socioeconomic status. The outcome was a list that included trustworthiness, respect, responsibility, fairness, caring, and citizenship.

Unlike the values clarification movement of the 1970s, initiatives based on the work of Kidder and Josephson make clear distinctions between right and wrong. Moral principles are distinguished from personal opinions. A character education framework is built on endorsing those values that civilized people agree on and encouraging thoughtfulness when making ethical decisions. Cheating on payment of taxes, being dishonest about children's ages to receive a lower admission cost at a theater, stealing supplies from work, violating traffic laws, and any number of other actions can and should be clearly identified as wrong.

The case for ethics based on a universal set of moral values is not new. C. S. Lewis (1952), a prominent twentieth-century Christian theologian, described the moral law or law of human nature and presented a strong case for recognition of this law by human beings everywhere. Lewis presented a compelling argument that universal moral laws were established by God, and humans were created with an inherent understanding of these ethical principles.

Ironically, vocal critics of ethics instruction in school settings often are members of the religious community who do not want to entrust discourse in this area to outsiders. Educators should equip themselves to present evidence of universal endorsement of ethical values being taught, but should avoid criticism of those who do not agree that this is an adequate rationale. For example, Christians might argue that agreement by many different people groups on a list of moral values is not a valid substitute for the authority of God. Having awareness and being able to acknowledge that the universal values being presented are also endorsed by the Word of God is a better strategy for defending ethics instruction than entering into arguments about religious belief systems.

Work ethic, historically associated with the Protestant ethic, is a construct that is closely related to ethics. Consisting of initiative, interpersonal skills, and dependability, work ethic encompasses providing an honest return for wages earned (Hill and Petty, 1995). The Protestant ethic was a term coined by Max Weber (1904, 1905) to describe characteristics of the hardworking Protestant people groups who were early settlers in the United States. Weber ascribed much of the success of capitalism to the work ethic of those pioneers. He also associated their work attitudes with their religious belief systems.

The list of attributes related to ethics recommended for emphasis in schools and other instructional settings includes integrity, responsibility, fairness, caring, initiative, interpersonal skills, and dependability. This list was derived from those studies that have identified moral values that are universally acceptable as well as from the literature on work ethic that focuses on attributes necessary for success in a technological world. Numerous additional descriptors could be added to the list, but a concise list is often more useful than an exhaustive list.

## THE ROLE OF ETHICS IN TECHNOLOGY EDUCATION

Ethics and ethical decision making have become increasingly important as technology has permeated the workplace (Hill and Womble, 1997). Technology has created an environment in which many people work in an autonomous environment, using wireless communications networks and portable information systems equipment to transact business. If these people lack ethical principles and a strong work ethic, productivity is impacted and economic vitality is impaired.

One of the distinguishing features of technology education is the extent to which it has encompassed a study of the impact of technology on society and culture, extending well beyond technical knowledge and skills. Understanding the interaction between technology and ethics is a viable topic. Technologically literate citizens should be cognizant of the ways technology has elevated the importance of ethical behavior in both the workplace and in private life.

Ethics have been a part of the study of technology from its inception. Character development was an integral part of many of the educational activities found in the historical movements that were precursors of technology education. John Locke, in 1697, emphasized the importance of virtue, wisdom, and manners as components of manual arts education (Bennett, 1926). Pestalozzi, Froebel, Saloman, Della Vos, and other philosophers and educators who contributed to the historical roots of technology education all gave considerable attention to character building and moral development (Scott and Sarkees-Wircenski, 2001).

In more recent years, ethics have had a prominent place in technology education initiatives. The *Jackson's Mill Industrial Arts Curriculum Theory* (Snyder and Hales, 1981) had a significant role in shaping technology education in its present form. The Human Adaptive Systems model presented in that document clearly recognized the interrelationship between ideological, sociological, and technological systems. Technology education content was organized around communication, construction, manufacturing, and transportation, but a holistic approach to study of these areas was endorsed. Particularly in the areas of technological impacts on individuals, society, and the environment, opportunities to consider ethical issues were to be an integral part of technology education programs.

The International Technology Education Association (ITEA) published the *Standards for Technological Literacy: Content for the Study of Technology* in 2000. This document, intended to guide and shape the future course of technology education, specifically included ethics as a component of technological literacy. Particularly in the area of technology and society, ethical considerations were specified as an important component of development, selection, and use of technologies. The recognition that technology could have both good and bad outcomes was described as a desirable outcome for technology education.

Technology education is a study of technology, which provides an opportunity for students to learn about the processes and knowledge related to technology that are needed to solve problems and extend human capabilities (ITEA, 2002). In almost every instance, one or more ethical issues can be identified in any situation in which technology is used to solve problems or extend human capabilities. Whether dealing with topics related to resources and the environment or deciding how a new medical technology should be implemented, ethical issues are often an important element in the technology education curriculum.

Two key factors in teaching ethics and values in technology education are being certain students understand the significance of ethics in a technological world and that they develop ethical decision-making skills. Technology education content should include integrity, responsibility, fairness, caring, and work ethic attributes of initiative, interpersonal skills, and dependability. This list represents core values that have been identified and widely endorsed as universally acceptable. It also includes characteristics relevant to success in a technological workplace identified through extensive research related to work in a technological world (Hill and Petty, 1995).

According to Kohlberg (1975), effective instruction related to ethics requires an approach that stimulates active thinking. Kohlberg identified this principle as a part of the cognitive-developmental approach described by Dewey. Hill and Womble (1997) identified an effective instructional design for teaching work ethic that provided an active role for learners and included several case studies, numerous small group discussions, and guided students to examine and reflect on their own attitudes toward work. Research has shown that ethics and related topics are not adequately addressed by lecture or other approaches that do not use active learning.

Technology education provides an ideal context for ethics instruction. The field has a long history of providing learning experiences involving engaged, hands-on instructional approaches. By incorporating ethical issues with content related to technological problem solving and development, students are provided opportunities to develop understanding that moves beyond technical skills and superficial knowledge of technical systems. Technology education that is consistent with the philosophical foundation expressed in the Standards (ITEA, 2000) and other seminal works must include these kinds of experiences.

The extent to which ethics instruction is incorporated in technology education varies depending on the instructor and the curriculum materials being used. Although developing character and encouraging ethical behavior is listed as an overall objective by most school systems, the approach to teaching ethics is often haphazard and not well-designed. Ethical issues are often embedded as a component of discussion points in technology education materials, but clear, focused strategies for including ethics in instruction is missing in many technology education programs.

## STRATEGIES FOR ETHICS INSTRUCTION

The extent to which practitioners teaching technology education incorporate ethics instruction in coursework is heavily influenced by the preparation provided by teacher education programs. Although professional ethics have been included in most technology teacher preparation programs, strategies for teaching ethics to students have not been included as a central theme of most technology teacher education programs. There are several reasons for this.

In most western world colleges and universities, ethical relativism has been a prominent philosophy during the past decades within disciplines related to teacher education. This resulted in a pervasive view that truth was relative, that objectivity and universal values were fantasies, and decision making was a subjective process (Kinnier, Dautheribes, and Therese, 2000). Promotion of certain values as universal was viewed as the disguised promotion of the dominant culture's values. The multicultural movement, by virtue of its idealization of diversity, also opposed promulgation of a universal set of values. In summary, the dominant position evidenced by most teacher preparation programs with respect to ethics has been that no individual or group of individuals is qualified to determine what is good or correct for all people. Therefore, ethics instruction has been almost entirely omitted from the curriculum.

Baker (1997) has provided a framework for using Kidder's ethical decision-making model in media instruction that could be adapted for use in technology education. Overlaying the revised Kidder model on a five-stage, problem-solving process familiar to technology educators yields the model shown in Figure 1-1.

In the first stage of the process described by this model, the moral issue is recognized, and persons with power to act are identified. The ability to see that there is a moral issue involved in a situation is important. In some circumstances, ethical issues are masked by cultural norms or peer behavior. Whether dealing with issues such as obtaining music using peer-to-peer Internet file-sharing software or making a decision about paying taxes on cash income, embedded ethical issues might be present but overlooked. Exposing learners to case studies and scenarios in which they are asked to apply the ethical decision-making model can increase sensitivity and raise awareness of ethical issues.

Learners also need to be able to identify who has the power to act in any situation involving an ethical decision. In the example of sharing music files, the person deciding whether to download a song from the Internet has a different level of empowerment than the person riding in a car when a CD containing downloaded music is played. Both might receive a benefit provided by unethical behavior, but the former had greater power to act than the latter.

Introduction to Ethical Issues in a Technological World

Figure 1-1. Revised Kidder Model for ethical decision making in technology education (Baker, 1997; Kidder, 1996).

| Define the Problem | ID Concepts and Ideas | Perform Relevant Tests | Develop/Test Prototypes | Produce Solution |
|---|---|---|---|---|
| Recognize that there is a moral issue.<br><br>Determine who has the power to act. | Gather the relevant facts.<br><br>Identify all possible options. | Is this a right versus wrong issue?<br>• Is one of the choices illegal?<br>• Is one of the choices intuitively wrong?<br>• Is there a choice that would be embarrassing if made public?<br>• Is there a choice that a very good person would not choose?<br>If an affirmative answer is given on one or more, make the right decision and move on.<br><br>Identify paradigm for right versus right issue.<br>Honesty* vs. loyalty<br>Individual vs. community*<br>Short-term vs. long-term*<br>Justice vs. compassion* | Apply resolution principles.<br><br>Utilitarian decision<br><br>Deontological decision<br><br>Justice-based decision<br><br>Virtue-based decision | Make the decision and implement respective action.<br><br>Ethical evaluation and reflection |

* Kidder identified four dilemma paradigms that are usually a part of right versus right dilemmas. The values identified with asterisks are those Kidder would recommend choosing in situations in which a pair of competing values appear to have equal weight.

The second stage of the model is the phase in which relevant facts are gathered and options are considered. Much like any problem-solving activity, this step in the process might involve brainstorming and research activities. If all of the facts are not identified, a flawed decision might result, with accompanying negative impacts. Learners need to also become adept in identifying options, particularly those that might differ from the norm. Some decisions might involve only the option to act or not to act. Others might have a complex array of options.

The model for making ethical decisions includes four questions to be asked in the third stage to determine the nature of the ethical decision. If any of these questions are answered in the affirmative, then the problem represents a moral temptation, and a right decision should be made or encouraged. When negative responses can be honestly given for each of the four questions in stage three, a right-versus-right dilemma is evident, and a set of four paradigms should be considered (Kidder, 1996). The dichotomies of honesty versus loyalty, individual versus community, short-term versus long-term, and justice versus compassion bring into focus the conflicts between core values that are present in the decision. In cases in which nothing in the situation compels a person to either of the competing values, Kidder recommended making a decision based on a preference for honesty, community, long-term interests, and compassion. These core values are identified in the model with asterisks to indicate their priority.

In the fourth stage of the decision-making model, philosophical positions are considered. In some instances, an affirmative answer to one of the questions provided in the third stage of the model provides guidance toward an ethical decision. In other cases, a right-versus-right dilemma is presented, and a difficult choice is needed. In either case, considering the philosophical perspectives listed in the fourth stage are of value. Identifying one's own framework for making ethical decisions within the range of possible positions can be a valuable exercise. Learning more about the resolution principles related to each perspective should be a part of ethics instruction. Self-reflection can then guide students to be more aware and consistent in their ethical decision making.

Earlier in this chapter, the philosophies of utilitarianism, deontology, justice ethics, and virtue ethics were presented. Technology teachers are usually out of their element in trying to teach or explain the nuances of these ethical frameworks, but if other teachers within a school are addressing these topics in humanities or social studies classes, opportunities for collaboration might be available.

Another way to help students understand the lines of reasoning represented in this stage of the model is to have them research materials related to Mill's principle of utility, Kant's categorical imperative, Rawl's theory of justice, and Aristotle's golden mean. Library or Web searches using these terms should provide an abundance of materials related to each of the ethical philosophies listed in the model's fourth stage.

Stage five of the model involves making and implementing the decision and evaluating the results. Any problem-solving model needs to have a feedback loop, and, in this instance, the consequences of decisions are reviewed to inform future decisions. When case studies or scenarios are developed for instructional use, consideration of this stage is important. Often, a case is described, but no effort is made to develop descriptions of possible consequences. It is important for students to be encouraged to think about what happens after an ethical decision is made so that they can evaluate the quality of the decision-making process.

Whether the model proposed here or some other model is used, a systematic approach for making ethical decisions is needed as a component of ethics instruction that is included within technology education. Technology teachers should be knowledgeable about a structured approach for making ethical decisions. Teachers can be prepared to provide quality instruction in ethics, whether through preservice teacher education or professional development experiences.

Along with an ethical decision-making model, technology teachers should be equipped to address stages of moral development. Familiarity with Kohlberg's (1975) model and awareness of one's own level of development is instrumental in the reflection and active thinking needed for effective learning to occur in the area of ethics. Instructional activities should include opportunities for students to learn about their own levels of ethical development.

If technology education is to address the ethical elements of the national standards and to successfully contribute to the technological literacy of students, it is important to provide opportunities for learners to develop ethical decision-making skills. This requires a conscious and deliberate effort. Pressures to cover specified technological content and other parameters placed on teachers by educational settings require that ethics instruction be integrated into existing course materials. There is not room for separate lessons and, even if there was, that would not be the most effective approach. Using case studies, group discussion, and other active learning strategies, students should be guided to consider ethical dilemmas within the context of various technologies being studied. Ethical decision-making strategies should also be taught and opportunities for reflection and practice need to be provided in conjunction with technology education technical content.

## REFLECTION QUESTIONS

1. What are some characteristics of a right versus wrong decision and how should choices in those situations be made?
2. How can a person determine when a right versus right dilemma has presented itself? Describe an example.
3. How do your own ethics and values compare with those of your parents or grandparents?
4. What have been the major influences shaping your own ethics and values?
5. When faced with an ethical dilemma, how can a person be consistent in making good decisions and upholding acceptable ethical standards?
6. What limitations do high school and middle school technology teachers face when considering how to include ethics instruction within their classes?
7. How would you respond to critics who argue that schools should not be involved in teaching children ethics and values?

# REFERENCES

Baker, S. (1997). Applying Kidder's ethical decision-making checklist to media ethics. *Journal of Mass Media Ethics, 12*(4), 197-210.
Bennett, C. A. (1926). *History of manual and industrial education up to 1870.* Peoria, IL: Charles A. Bennett.
Gilligan, C. (1982). *In a different voice.* Cambridge, MA: Harvard.
Goree, K. (1996). *Ethics in American life.* Cincinnati, OH: South-Western.
Hill, R. B., & Petty, G. C. (1995). A new look at employability skills: A factor analysis of the occupational work ethic. *Journal of Vocational Education Research, 20*(4), 59-73.
Hill, R. B., & Womble, M. N. (1997). Teaching work ethic: Evaluation of a 10-day unit of instruction on work ethic, work attitudes, and employability skills. *The Journal of Educational Opportunity, 16*(1), 57-79.
International Technology Education Association (ITEA). (2000). *Standards for technological literacy: Content for the study of technology.* Reston, VA: Author.
International Technology Education Association (ITEA). (2002). *Teach technology.* Retrieved September 24, 2002, from http://www.iteawww.org/TeachTechnology/whatistechteaching.html
Johnson, D. G. (2001). *Computer ethics* (3rd ed.). Upper Saddle River, NJ: Prentice-Hall.
Kidder, R. M. (1996). *How good people make tough choices.* New York: Simon & Schuster.
Kidder, R. M. (1994, July/August). Universal human values: Finding an ethical common ground. *The Futurist,* pp. 8-13.
Kinnier, R. T., Dautheribes, J. L., & Therese, M. (2000). A short list of universal moral values. *Counseling and Values, 45*(1), 4-17.
Kohlberg, L. (1975). The cognitive-developmental approach to moral education. *Phi Delta Kappan, 14,* 670-677.
Lewis, C. S. (1952). *Mere Christianity.* New York: Macmillan.
Maley, D. (1973). *The Maryland plan.* New York: Benziger Bruce & Glencoe.
*Merriam-Webster's Collegiate Dictionary.* (1998). Retrieved September 13, 2002, from http://www.m-w.com

Nish, S. (Ed.). (1996). *Good ideas to help kids develop good character.* Marina del Ray, CA: Josephson Institute of Ethics.

Scott, J. L., & Sarkees-Wircenski, M. (2001). *Overview of career and technical education.* Homewood, IL: American Technical Publishers.

Secretary's Commission on Achieving Necessary Skills (SCANS). (1992). *Learning a living: A blueprint for high performance, A SCANS report for America 2000.* Washington, DC: U.S. Department of Labor.

Snyder, J. F., & Hales, J. A. (Eds.). (1981). *Jackson's mill industrial arts curriculum theory.* Charleston: West Virginia Department of Education.

Tucker, E. M., & Stout, D. A. (1991). Teaching ethics: The moral development of educators. *Journal of Mass Media Ethics, 14*(2), 107-118.

Vincent, A., & Meche, M. (2001). Use of ethical dilemmas to contribute to the knowledge and behavior of high school students. *High School Journal, 84*(4), 50-58.

Weber, M. (1904, 1905). *The protestant ethic and the spirit of capitalism.* Translated by T. Parsons. New York: Charles Scribner's Sons.

# Ethics in a Culturally Diverse Technological World

Chapter 2

Linda Rae Markert
State University of New York at Oswego
Oswego, NY

> Professional ethics courses make minor contributions to major needs . . . real world ethics is a complex admixture of personal, social, and professional morality.
> Robert J. Nash, 2002, Real World Ethics

> If we are to understand the nature of ethics in a diverse workforce, we must first learn something about the different streams of ethical thought and examine them within the context of diversity.
> Willie E. Hopkins, 1997, Ethical Dimensions of Diversity

It is nearly impossible to get through the day without hearing words like global, worldwide, or international at least once and likely several times. Finding a remote place of respite devoid of any connectivity to civilization is similarly difficult for any of us to achieve for any significant period of time. Countless artifacts of science and technology surround us and greatly influence our interactions with other persons on a daily basis. Middle school students routinely (sometimes too often!) have online conversations with individuals around the world whom they may never have met face-to-face, and they believe this is quite normal. High school students are learning to use sophisticated handheld Global Positioning System (GPS) devices in their science classrooms, establishing local connections to distant geostationary satellites and illustrating the outer limits of global systems that even extend to outer space.

Many of us take these daily occurrences for granted, lending further credence to the often-heard assertion that we live in a global society that is intensely interconnected and is, therefore, interdependent. On the other hand, the number of persons among us who have spent extended periods of time away from the United States in an attempt to live among and learn from persons in other countries, is certainly a minority. Stated another way, all of us may be aware that our world is culturally and technologically diverse, but very few of us are able to fully comprehend and articulate what life is really like and about in other places away from our homeland.

The purpose of this chapter centers on an examination of the extent to which ethical values influence the development and transfer of technology around the world. As noted in Chapter 1, ethical values include such qualities as integrity, responsibility, fairness, caring, and a dedicated work ethic. The globalization of scientific and technological research and education has created a complex network of partnerships, linkages, joint ventures, and numerous multinational enterprises. Coupled with this reality, the current strong position of the United States as the world's leading producer of high-technology products reflects its success in attending to the needs and desires of a large domestic market, as well as in responding to the demands of foreign markets (National Science Board 2000). In considering the importance of ethics and ethical behavior as a facet of the new technology education standards, the following sections of this chapter provide a brief overview of our own nation's ethical and moral underpinnings, several international perspectives on ethics, and the role ethics in society has played in the development and use of technology in our world.

## ETHICS IN THE UNITED STATES OF AMERICA

The term culture, from the Latin word *cultura*, can be used to refer to the amalgam of socially transmitted behavior patterns, beliefs, and other products of human thought that are characteristic of a population. McElroy (1999) suggested that a historical culture can be viewed as a unique set of extremely simple beliefs that are formed and communicated through behavior over more than three generations. Cultural beliefs must be simple to make sense to many people and to be expressed in varying behaviors over an extended period of time. The concept of ethics is derived from the Greek word *ethos*, which Byron (1977) loosely translated to mean internal character. In many instances, as you learned in Chapter 1, the terms ethics and values are used interchangeably. However, as you also learned, distinctions can be made between them. Rokeach (1968) stated that ethics tend to focus on the conduct of individuals, whereas values represent the fundamental beliefs that individuals hold to be true about conduct. Stated differently, values are the underlying beliefs and attitudes that help determine one's actual conduct. Hopkins (1997) further explained that (*a*) a value establishes a moral standard for an individual such that action may be taken to achieve a goal, and (*b*) the purpose of an ethic is to ensure that the action designed to achieve a goal will be done without violating a value. The term morals, or the concept of morality, is also used as a synonym for ethics. Once again, *ethos* from the Greek language is translated into what may be called internal character. However, the Latin translation of *ethos* is *mos, moris*, from which the term moral is derived and seems to shift the focus from internal character to observable behaviors (for example, actions, habits, traditions, or customs). Figure 2-1 illustrates the relationships between culture, ethics, values, and morals.

Figure 2-1. Relationships between culture, ethics, values, and morals.

The dictionary defines teleology as the philosophical study of design or purpose in natural phenomenon. The Greek word *teleos*, meaning "end" or "issue," is the root word. Teleology is the underlying premise for utilitarianism, one of the ethical frameworks presented in Chapter 1. Hopkins (1997) asserted the following:

> Teleological theories of ethics hold that whether an act is morally right or wrong depends solely on how good or bad the consequences of the action are for oneself.... The teleological perspective on ethics argues that acts are morally right or good if they produce some desired state of goodness or pleasure and are morally wrong or bad if they produce some undesirable state of badness or pain. Subsequently, the rightness or wrongness of actions is determined by the results that these actions produce and not the act itself. (P. 26)

Historians of American culture grapple with this question: In our contemporary technological society, how is it that so many different people who have widely disparate, diverse, and often dissimilar ideas and customs are able to get along as Americans (McElroy, 1999; Wolfe, 2001)? The unity of this vast and diverse nation we know as the United States of America has evolved over many generations because common behaviors based upon strong principles/values have endured. The earliest experiences of the American colonists on the continent's Atlantic coastal plain region were nearly unbearable. These pioneers confronted a veritable wilderness where they had to devise ways to build communities and survive together in a harsh environment. Generations of individuals labored diligently to transform an expansive rough country and shaped the history of America. McElroy thus concluded that it truly was (and continues to be) work that shaped (shapes) the primary beliefs of the American people, because the most important task for generations of settlers was survival, and those who did not work could not survive.

If we can acknowledge McElroy's (1999) conclusion that the primary beliefs of American culture are directly related to work, we might be persuaded to agree with these simple tenets: (*a*) "everyone must work," (*b*) "people must benefit from their work," and (*c*) "manual work is respectable" (p. 37).

Our country was from the earliest beginning a society of workers, and it remains so to this day. Most of us learn when we are quite young that work is respectful, and we should endeavor to be successful and self-supporting. The beliefs about hard work that have developed over time in our culture make it feasible for an electrician in twenty-first-century America to earn a higher salary than a university professor and be an equal member of the middle class. The development of our civilization from a brutal wilderness made America, as Benjamin Franklin labeled it, a "Land of Labour" (cited in McElroy, 1999). Beyond our reverence for work, ethical behaviors in the United States are based largely on the principles of freedom, equality, individuality, responsibility, improvement, and practicality (McElroy).

## BELIEFS OF AMERICA'S FOUNDING FATHERS

Politically and socially active during the last quarter of the eighteenth and first decade of the nineteenth centuries (roughly 1774 to 1809), the Founding Fathers were the group of men who created the American Republic. Three major spheres of activity in which these men participated collectively were the American Revolution, the Constitutional Convention of 1787, and the establishment of the federal government in 1789 (Padover, 1960). Generally speaking, the Founding Fathers were solid citizens, who were respected in their community, and usually of good family background. "Of the signers of the Declaration of Independence, the Articles of Confederation, and the federal Constitution, nearly half were lawyers and at least fifteen were businessmen . . . five were physicians . . . and sixteen, among them the learned and brilliant James Madison, had no profession other than politics" (p. 28). The majority of these men were native born, almost entirely of British (including Irish) heritage.

The prevailing intellectual climate of their era was that of the eighteenth century Enlightenment—with its emphasis on reason and demonstrable scientific truth. This environment gave potency and philosophic meaning to the daily experiences of an emerging new man—the practical, sensible, down-to-earth, energetic doer and builder: the American. The social, political, and economic realities of the decades immediately following the American Revolution were a driving force for the creation of a strong, centralized government headed by a powerful chief executive. The Founding Fathers' ideas contained in Article II of the Constitution continue to shape the presidency in our country to this day (Bunch et al., 2000).

The spiritual world of America's Founding Fathers was predominately one of Protestantism. The Protestant roots were deep, even for those men who were not religious and who did not belong to any established church (Padover, 1960). They collectively displayed Calvinistic beliefs in their stubborn sense of personal independence and systematic refusal to accept authority without questioning it. John Adams is quoted as saying that the hatred of the Church of England "contributed as much as any other cause to arouse the people against Britain's political authority" (cited in Padover, 1960, p. 45).

This Protestant tradition of dissent eventually culminated with the Founding Fathers' establishment, first in principle and then in practice, of the separation of church and state. To the extent that spiritual liberty is the first of all personal liberties, the permanent disestablishment of the church from government was one of the Founding Fathers' greatest achievements. "An official religion, that is, a church for which the people are compelled to pay taxes regardless of their own beliefs, produced, the Founding Fathers felt, a chain of evils of which the foremost were the denial of the free exercise of reason and the perpetuation, through coercion, of moral hypocrisy" (Padover, 1960, p. 45). Take a brief moment to consider the following quotes, relative to the separation of church and state, written or expressed by four of our country's Founding Fathers:

> *George Washington (1732–1799).* We have abundant reason to rejoice that in this Land the light of truth and reason has triumphed over the power of bigotry and superstition, and that every person may here worship God according to the dictates of his own heart.

*John Adams (1735–1826).* The United States of America have exhibited, perhaps, the first example of government erected on the simple principles of nature; and if men are now sufficiently enlightened to disabuse themselves of artifice, imposture, hypocrisy, and superstition, they will consider this event as an era in their history.

*Thomas Jefferson (1743–1826).* Everyone must act according to the dictates of his own reason, and mine tells me that civil powers alone have been given to the President of the United States, and not authority to direct the religious exercises of his constituents.

*James Madison (1751–1836).* And I have no doubt that every new example will succeed, as every past one has done, in showing that religion and government will both exist in greater purity, the less they are mixed together.

(Words of our American Founding Fathers, 2002)

In summation, this nation's Founding Fathers were men of strong convictions, who were animated by a deep moral conscience and devoted to the ideal of freedom. Their world was ruled by reason, pragmatism, and the philosophy of natural laws. Still, their regard for personal spirituality was evident. Our American Republic was crafted by these men on their intrinsic belief that it was forever possible for people to govern themselves without abuse or injustice (Padover, 1960). Most assuredly, their views of the world resulted in a democratic system of government that ultimately became friendly to capitalism and technological development in our nation.

## THE PROTESTANT WORK ETHIC

Any discussion of ethics in the United States of America would be grossly incomplete without serious mention of what is commonly referred to as the Protestant Work Ethic (PWE). Generally speaking, the PWE is respected as a code of morals that are based on the principles of thrift, discipline, hard work, and individualism. The person to whom most credit is given for the formulation of the PWE is the German political philosopher and economic sociologist Max Weber (1904, 1905). He perceived and examined the close relationship between the Protestant ethic and the rise of capitalism.

Furnham (1990) suggested that Weber understood capitalism as a mass phenomenon, a culturally prescribed way of living, and a moral doctrine to advance individuals' material interests. Weber (1904, 1905) himself stated that capitalism is "the rational and calculated expectation of profit by the utilization of opportunities for exchange" (p. 22). Weber took great interest in the fact that capitalism had developed mainly in those areas of Europe in which Calvinistic Protestantism had taken a foothold early in the Protestant Reformation. In his quest to build a case that a causal relationship existed between religion and economic life, Weber's work gained much notoriety, and his thesis has survived as one of the best known and also quite controversial works in the social sciences.

Persons who belonged to the Protestant faiths were intensely anxious about their state of grace with God, largely due to the doctrine of predestination, which is central to Calvinism. Those who believe in predestination realize that God's grace is as impossible for those to whom he has granted it to lose, as it is unattainable for those to whom he has denied it. Therefore, persons spent considerable time worrying about whether they were one of the elect and certain of everlasting life. A practical means of reducing this anxiety took the form of a systematic commitment to a "calling"—that is, to hard work, thrift, and self-discipline. Material rewards recouped as a result of this work were to be saved and reinvested. Ultimate success in the commercial world tended to have a reassuring effect for individuals because they believed they were safely in God's good grace. Stated another way, persons who worked hard, practiced frugality, and were outwardly successful could be assured of a blessed afterlife.

Oates (1971) seems to concur with this summation in his interpretation of Weber's monumental work. He makes the following statement about the PWE:

> The so called Protestant Work Ethic can be summarized as follows: a universal taboo is placed on *idleness,* and *industriousness* is considered a religious ideal; *waste* is a vice, and *frugality* a virtue; *complacency* and *failure* are outlawed, and *ambition* and *success* are taken as sure signs of God's favour; the universal sign of sin is *poverty,* and the crowning sign of God's favour is *wealth*. (p. 84)

Without question, the idea of the Protestant ethic has significantly influenced our nation's history, sociology, and political science.

Throughout the last quarter of the twentieth century, however, ethics have become more secularized and seem to be less tied to spiritual convictions.

Several examples of contemporary ethics were discussed in the research completed by Maccoby and Terzi (1979). They argued that the Puritan ethic, craft ethic, entrepreneurial ethic, and career ethic have developed sequentially in America and are all related to one another. The career ethic, perhaps most evident in contemporary society, "emphasizes meritocracy, talent, and hard work within organizations leading to success and promotion. This ethic implies an other-directed, ambitious, marketing character" (Furnham, 1990, p. 15). As you can see, there remain vestiges of the Protestant ethic, but they are more or less devoid of the salient link to God's grace and the life hereafter.

Another ethical philosophy that emphasizes a worldview based on natural phenomena is labeled secular humanism, a term that has come into use in the last 30 years. Secular humanists accept a philosophy called naturalism, in which the physical laws of the universe are not superseded by supreme beings, such as demons, gods, or other spiritual entities outside the realm of the natural world (Kurtz, 2002). They view ethics as an autonomous field of inquiry, independent of theological claims, amenable to rational scrutiny, and espouse testing value judgments by their consequences.

Although secular humanism is apparently at odds with faith-based religious dogma on many issues, its proponents state it is dedicated to the fulfillment of the individual and humankind in general (Council for Secular Humanism, 2002). In sum, secular humanists do not rely upon gods or other supernatural forces to solve their problems or provide guidance for their behavior. Therefore, secular humanists do not believe in God or an afterlife. Secular humanism encourages people to think for themselves and question authority, and suggests that the morality of our actions should be judged by their consequences in this world (Cherry and Matsumura, 1998).

To a large degree, the backdrop of the Founding Fathers, the PWE with endorsement of hard work, and the influences of secular humanism explain the prominence of utilitarianism in the American legislative and judicial systems. The principle of basing decisions on the greatest amount of good for the largest number of people is very pragmatic and allowed people with diverse belief systems to successfully establish a democratic system of governance.

## IMMIGRATION AND AMERICAN ETHICS

*E Pluribus Unum* is our national motto, meaning "from many, one." It was originally conceived to describe and celebrate the unification of 13 states into one union. Throughout our country's history, this phrase has often been used to exemplify the fact that the vital and vibrant unity of our much larger national community is founded on individual freedom and the diversity that emanates from it.

More than 55 million immigrants have made the choice to leave their homelands and resettle in America over the last four centuries. This fact represents the largest movement (most often voluntary) of human beings to any one place in the history of mankind (McElroy, 1999). Restrictions on immigration to the United States were not enacted to any significant legislative extent prior to the early twentieth century. Contemporary immigration policy in our country is written in the Immigration Act of 1990 (P.L. 101–649), and a considerable volume of migration to America continues as we enter the twenty-first century. Historically, most Americans have either been immigrants themselves or descendants of immigrants; therefore, the beliefs and behaviors of persons from many foreign lands have had a fundamental and determining influence on the formation of American culture, ethics, values, and morals.

The long-term effects of immigration are complex, numerous, and diverse. Immigrants contribute to the social fabric of the United States in countless ways: (*a*) to its vibrant and diverse communities, (*b*) to its lively and participatory democracy, (*c*) to its vital intellectual and cultural activities, (*d*) to its renowned job-creating entrepreneurship and competitive marketplaces, and (*e*) to its family values and strong work ethic (U.S. Commission on Immigration Reform [USCIR], 1997). Our current policies regarding immigration are regulated such that priorities for admission are established, reunification of nuclear families is facilitated, U.S. employers are given access to a global labor market while ensuring that current U.S. workers are not displaced, and we are able to fulfill our commitment to principles of humanitarian protection and assist in the resettlement of refugees.

Immigrants to the United States prior to the 1800s were almost entirely from Europe; since that time, increasing numbers of persons from Asia, Africa, and Latin America have immigrated to this expansive nation. Today, the majority of immigrants being admitted represent cultural groups from Asia, the Caribbean, Mexico, and Central and South America (Hopkins, 1997). Regardless of their country of origin or the century during which they came, these millions of immigrants traveled to America in search of a better and much improved life for themselves and their families. McElroy (1999) surmised that these diverse groups of immigrants brought with them three simple beliefs, which their behaviors and decisions as resettled citizens demonstrated: (*a*) "improvement is possible," (*b*) "opportunities must be imagined," and (*c*) "freedom of movement is needed for success" (p. 61).

Ostensibly, if persons who migrated to a new land to start a new life possessed these beliefs prior to their arrival, they would likely be prepared to adjust to the social and ethical beliefs of the new land. These three beliefs represent a framework that allowed diversities of nationality, language, and religion to be gradually amalgamated into a new American identity (McElroy, 1999). Throughout our history, each immigrant and immigrant group has had a unique set of experiences in America, distinctive to the individual person or group. The process of becoming an American is most simply called "Americanization" and it entails personal choices and meaningful decisions.

There exists no mandate that everyone who immigrates to the United States must dismiss the customs and practices brought with them from their homelands. The fact that most immigrants hold onto their native customs to some extent actually makes the face of our nation all the more intriguing. On the other hand, according to classical assimilationist theory, the best option for newcomers to a given society is to shed their ethnicity as quickly as possible (Hopkins, 1997). Evidence exists that most immigrants "choose an option that is somewhat less drastic, and they either (*a*) assimilate the mainstream's cultural values, (*b*) assimilate a particular minority's or subculture's values, or (*c*) preserve their own cultural values" (p. 58).

The concept of ethnic culture has developed alongside the continuing waves of immigration over the years. This idea illustrates a component of ethnicity that refers to a pattern of unique behaviors, beliefs, and ethics that sets a cultural group apart from others (for example, Native Americans, Hispanic Americans, Asian Americans, and African Americans). The process of Americanization and patterns of cultural differences can be studied via the lenses of both assimilation (as previously noted) and deculturation. Under assimilation, the contention is that members of ethnic cultures (immigrants) adapt their behavioral patterns, values, and norms to those of the dominant culture. In so doing, these individuals may camouflage their true feelings and actually suppress aspects of their own culture while in public. Under deculturation, members of ethnic cultures (immigrants) retain their distinct set of norms and values with no attempt to integrate or synthesize the value system (ethics, morals) of the dominant culture (Hopkins, 1997). A primary example of deculturation is the presence of a vibrant "Chinatown" in the middle of any large American city, where there is minimal interaction between the residents of Chinatown and persons residing outside that small community.

Regardless of the process used to become an American, immigrants have exposed this nation to a wide array of cultures and have built our undisputed reputation as a conglomeration of people from around the world. Immigration is a central theme in the story of the United States of America. The beliefs of our Founding Fathers underlie our core values and have formed our ethics and morals. Through the years, millions of immigrants have subscribed to a good work ethic, strong family values, and a belief in freedom and justice for all citizens. It is important for us today to recognize that although we are a nation as one whole, it is really the unique and diverse individuals who make the whole so strong and appealing to others who reside outside our land (*E Pluribus Unum*). The following section of this chapter presents an array of international perspectives on ethical behavior and describes some of the external perceptions different ethnic groups have about persons who call themselves Americans.

## INTERNATIONAL PERSPECTIVES ON ETHICS

The current state of world politics and international agreements give credence to the fact that cultural awareness is extremely important in contemporary society. As noted earlier, the number of American citizens who have spent extended periods of time living with and learning about persons from other cultures is rather small. Regardless, those Americans who are traveling abroad and experiencing other cultures and customs undeniably make an impression on those persons they meet on foreign soil. When diverse cultures meet, it is not uncommon for misunderstandings, misconceptions, and erroneous assumptions to arise. Because the United States is commonly viewed as the last remaining superpower, Americans are perhaps the most loved, hated, envied, appreciated, and resented persons on the planet. Perceptions such as these become evident when Americans visit other nations and when foreign visitors take time to explore the United States.

When persons of foreign nationalities visit the United States, they often notice the goodness of the average American citizen. The fact that immigrants perceive America as providing freedoms and opportunities being denied to them in their homeland makes the United States a highly desirable destination. Foreign students studying in American universities or high schools remark that their classmates are helpful and find their empathy, candor, humor, and hard work to be worthy of great notice. Hopkins (1997) suggested that "Americans are often viewed by members of other cultures as being very informal, direct, competitive, achievers, questioners, punctual, and obsessed with cleanliness," whereas "Americans view themselves as being caring and generous people who value their independence and entrepreneurial spirit" (p. 44).

Realistically, it is impossible to deny the existence of anti-American sentiments around the world, both now and in decades past. Hussain (2001) concluded that negativism about America has largely been derived and shaped by popular perceptions in three areas: (*a*) dignity, (*b*) double standards, and (*c*) democracy. He surmised that the aforementioned American goodness is not generally exported, remaining principally confined to its homeland shores. He further stated that the ethics and values Americans purport to be true at home—liberties, rule of law, and democracy—are rarely exemplified in American foreign policy (Hussain). Graham Fuller, who is former vice chairman of the National Intelligence Consulate of the Central Intelligence Agency, in an interview before the U.S. Department of State, stated the following:

> There is a huge cadre in the Middle East of people educated in the United States at the university level who have the warmest and fondest memories of this country. You can meet hundreds of them at any gathering. They will tell you about their time all over this country, in the Midwest the hospitality they encountered, their admiration for Americans' political values for democracy, for human rights, for minority rights, this kind of thing. But they say, "We do not recognize your country when we see your policies in our part of the world. We don't see these American values reflected at all." (Commission Reviews Middle Eastern Perceptions of the United States, 2002)

The horrific terrorist attacks on the United States in September 2001 drew renewed international attention to the concept of anti-Americanism. The deliberate strikes against deeply symbolic and valued objects in America terrified observers and citizens by the sheer force of hatred directed at the United States (Gudkov, 2002). Dislike of America is consistent with the common irritation an impoverished civilization experiences toward a wealthy and powerful neighbor nation. In many ways, the incredible, and perhaps intolerable, success of the United States during the twentieth century made it the technological and economic leader of the world. At the same time, persons from other nations, citizens and leaders alike, experienced grating envious reactions characterized by the question, "Why not us?"

We live in an interconnected society wherein the continuum of technological prowess and development is expansive. Behaviors and decisions with regard to foreign relations and agreements may not always be ethical by one's personal definition, however much we hope they should or will be. Perspectives of leaders in other nations relative to ethics are important to understand and acknowledge as we enter into numerous exchanges involving the transfer of science and technology.

## *Ethics in Africa*

Around the world, we can find examples of great concern regarding ethics and corruption. Leaders in some regions are particularly interested in finding ways to mitigate the damaging effects unethical and dishonest practices have had on the technological and economic development of their countries. Africa currently faces enormous challenges in its efforts to achieve sustainable human development. This continent is home to many of the world's poorest countries and it is overwrought with endemic diseases, such as malaria and HIV/AIDS. A significant percentage of Africans also live in countries experiencing severe civil conflict and unrest (United Nations Department of Economics and Social Affairs Division for Public Economics and Public Administration [UN/DESA/DPEPA], 2001).

There are certainly pockets of gain and improvement in various parts of this vast continent in the world's southern hemisphere, but overall, prospects for development are not very promising. With declining export shares of primary commodities, a lack of viable manufacturing and service industries, along with capital flight and brain drain, Africa's position in the global economy continues to falter (UN/DESA/DPEPA, 2001). The paramount preconditions to Africa's capacity for sustainable development appear to involve improving governance, resolving conflict, and attending to the critical medical and health needs of the populace.

The nature of human existence in Africa is communal, and it is a reality that heads of households, officers of spiritual associations, and various religious specialists are viewed as moral guardians of their families, groups, and society at large. They are regarded as guardians of the moral order of the universe through their observance and transmission of both life and tradition. Similarly responsible on a substantially higher level are Africa's kings, chiefs, clans' leaders, and other types of authority figures whose power extends beyond the family or the small community.

Magesa (1997) explained that the concepts of law and politics are ambiguous when applied to African organizations because they cannot be easily extracted from the religious moral and ethical systems. Stated another way, in traditional African countries, there is generally no specific political structure that is distinct from the social and religious organization in society. When persons occupy political and religious positions of some importance in Africa, their political power is visibly reinforced at those points in the seasonal cycle, or the group's developmental cycle, where ritual officership gives them enhanced authority (for example, during a religious day of atonement; Turner, 1966). African leaders who are at the higher social levels of the lineage, clan, or ethnic group personify the order of the world and the harmony that enables life to continue. They believe their primary purpose as leaders is to guard the power of life in the community (Magesa).

Law and resolution of conflict in African societies are integrally related to the entire system of morality and ethics practiced in African religions. Gluckman (1965) asserted it is difficult to separate law (governance) from custom, taboos, divination, mediumship, ordeals, and the expectations of sharing, harmony, play, and good company in general. Magesa (1997) elaborated on this notion of good company stating it "implies community, that is, the establishment and maintenance of harmonious relationships among people . . . and includes the exchange of aid and sympathy which spring from personal friendship" (p. 259). Good company, practical sharing, communion, and communication are essential factors of African political systems. The African legal system and moral code of conduct are inextricably linked, and resolution of conflict is commonly connected to religious practices and beliefs.

African religion emphasizes the communal nature of property within a given community reflecting the principle of inclusion. It does not dismiss private or personal ownership, but the ethical task is to establish a balance between the rights to private ownership of property and the human understanding of the resources of the universe (Magesa, 1997). In African religious thought, the right of personal ownership resides within the context of joint or public right of access to the basic resources that are essential for life. For example, when any form of tribute is given to the chief or other leader of the community (for example, cattle, grain, water, or labor), it is paid to them in trust for the entire community. Africans believe that there are some resources that are gifts from God to all human beings and, therefore, cannot be privately owned. Land is a primary example which Africans regard as an absolute source of sustenance, and it may only be held in trust. Society entrusts pieces of land to individuals or groups for their own use but also for the greater society's well-being, growth, and development (Magesa).

In summation, the African religious worldview emphasizes relationships and centers on the fact of creation. God, through the act of creation, is omnipresent in the entire universe. Humanity, at the center of the universe, is firmly connected to all living and nonliving creations by means of each creature's life force. Africans believe that God, spiritual beings, ancestors, humanity, living things, and nonliving things possess life forces with greater and lesser powers, and all forces are intertwined (Magesa, 1997, p. 285). From birth to death, African religion pays special attention to all the rights of passage marking different stages of development of these life forces. Economic activities and political agreements, by their relation to life itself, come to be viewed as religious events.

*Ethics in Asia*

Similar to what we find in African nations, a common thread woven through Asian cultures is the strong influence of dominant religions, such as Confucianism, Taoism, and others in the countries aligned with China's ideals, and of Buddhism and Shintoism in Japan. A value emphasized in these religions is social interaction, which Hopkins (1997) suggested is the basis of the strong group identification, formality and courtesy, humility, and taciturn demeanor for which Asian cultures seem to be known.

In Asian business organizations, the prevailing values highlighted by corporate culture are loyalty, accommodation, and honoring authority.

One example is exemplified in Korea's family-run conglomerate. The founder of Samsung wrote an employee policy in 1938 explaining that loyalty to the organization would be highly valued in all workers (Andrew 1988). Policies like this are evident in many Asian firms, and, in terms of the accommodation ethic, Asian employees generally try to minimize dissent and avoid confrontation. Asian organizations tend to be very hierarchical, and the chain of command is clearly defined and respected. Employees who are working in the lower levels of the company are extremely careful to not offend individuals of higher authority or of greater age.

Enderle (2000) suggested that Confucian ethics are the spiritual source for and exist at the heart of the economic successes experienced in the Asian countries of Taiwan, South Korea, Hong Kong, and Singapore. Cua (1992) explained that Confucian ethics is a form of virtue ethics. The goal is a well-ordered society based on good government that is responsive to the needs of people, to issues of wise management of natural resources, and to just distribution of burdens and benefits (Enderle). *Tao* is the ethical ideal of a good human life that underlies Confucian ethics and stresses character and personal formation or cultivation of virtues. These virtues consist of (*a*) the basic independent virtues of love and care for one's fellows, a set of rules of proper conduct, and reasoned judgment concerning the right thing to do; and (*b*) the dependent virtues of filiality, respectfulness, and trustworthiness (Cua).

In Japan, there are some cultural values and ethical behaviors that are not replicated in other Asian cultures (Hopkins, 1997). The influence of Confucianism can be seen in their emphasis on hierarchy and position, whereas the subtle and indirect demeanor and hidden meaning in the Japanese disposition is linked to Zen Buddhism. Japanese persons view themselves as members of a group first and then as individuals, and they do not enjoy being singled out to be praised or congratulated. In Japanese organizations, employees are expected to get along in a group environment, to adhere to the established formalities, to respect the clear class distinctions, and to behave in a conventionally predictable manner (Goodman, 1990).

The cultural value system and ethics are dominated by the qualities of sameness, evenness, and consistency. Corporations doing business in Japan foster norms that focus on courtesy, conformity, and caring for others. Harmony is of paramount importance (Hopkins, 1997). Deference and respect are based on age, rank, role, and gender. The suppression of women in these corporate settings is another salient factor that characterizes the value system of Japanese organizations.

The Republic of India, a sprawling land of contrasts and seeming contradictions, is home to a large number of diverse ethnic communities. Since becoming independent in the late 1940s, the Indian government's basic philosophy of development has been self-reliance. The dominant religion in the country is Hinduism, but substantial minority religions are practiced including Islam, Sikhism, Jainism, Buddhism, Christianity, and Judaism. Indian ethics and value systems do not assimilate these many different cultures, customs, beliefs, languages, and religions into a unified whole—each exists and is accepted as they are (Hopkins, 1997). Included among the values of the middle class are respect for education and competitive excellence. As in other Asian countries, there exists in India a deep regard for age and social position. One of the best known aspects of India's cultural structure is the caste system, which prescribes social status at birth and offers no opportunity for upward mobility.

Although employees in Indian business organizations appreciate the concepts of time and protocol, appointment schedules are not always strictly adhered to, and perspectives toward work are somewhat more relaxed than in most other Asian cultures (Singh and Hofstede, 1990). Formal titles are used almost all the time, but upper-level Indian managers seem generally inclined to use consultative and participatory styles of decision making. At the lower levels of the organizational chart, however, managers seem to be more autocratic, and sharing any power with one's subordinates is viewed as a weakness (Singh and Hofstede).

## *Ethics in Europe*

According to Swenson (2000), business ethics programs are now receiving greater attention in Europe. He noted that in the early 1980s less than 20 percent of the major corporations in Germany, France, and the United Kingdom had a published code of conduct. By the early 1990s, that population of firms had increased to the 50 percent mark. Swenson went on to say that codes of ethics are not necessarily a panacea for unethical behavior, but they represent a good starting point.

A large number of countries make up the geographic region of Western Europe. In contemporary society, it is apparent that the European

Union stands together with the United States in a deep-rooted commitment to promoting shared values of democracy, human rights, and fundamental freedoms in our world. Both parties promote the advancement of the common goals of peace, development, and prosperity. A strong foundation for substantial and influential dialogue on foreign policy issues is in place for today's political leaders in the European Union and the United States. The ethics and values of two Western European countries—France and the United Kingdom—are discussed briefly, to provide a partial profile of this region's perspective on ethics. As noted earlier in the discussion of ethics in the United States of America, many of the early immigrants to America emanated from these European nations. For this reason, we see similarities in their beliefs and value systems.

Generally speaking, persons in Western Europe tend to be rather formal and conservative, and commonly refrain from using an individual's first name without invitation (Hopkins, 1997). Punctuality is viewed as a sign of courtesy in this region of the world, and persons holding academic titles or degrees expect them to be used as signs of respect (for example, doctor, chancellor, dean, professor, rector).

The predominant religious background of most people in both France and the United Kingdom is Christian—France being mostly Roman Catholic and the United Kingdom primarily Anglican. The ethical commonalities between these affiliations include a respect for discipline and responsibility, a low tolerance for ambiguity, a view of oneself as an individual first and then as a member of a larger group, high mobility, and great esteem for formal education (Hopkins, 1997).

The ethnic population of France is a broad blend of many different groups—among them Celtic, Latin, Nordic, and North African Arabs. It is largely due to this diverse ethnic pool that French persons consider themselves members of a family first, then citizens of France, and finally members of organizations. The French culture exudes a flair for the arts, performance, and the joy of living. The concept of success in France is not linked to a person's direct accomplishments, but is commonly aligned with her or his educational level, family heritage, and financial status in society (Hopkins, 1997).

The ethnic mix evident in the United Kingdom is primarily English, Scottish, Irish, and Welsh, with considerable numbers of individuals from West India, East India, and Pakistan. Despite this diversity of cultures, Hopkins (1997) explained that a basic sense of fair play underlies this nation's ethics and value systems. The British seem to be an aloof group, but their demeanor is more a reflection of privacy and personal modesty coupled with a desire to not display too much emotion in public. The melding of democratic principles with a much adored monarchy in the United Kingdom has created a culture that is quite formal and conservative, but one that continues to value personal space and independence.

In European firms, a strong concept of social order and emphasis on rules are evident in the corporate value system (McLaughlin, 1990). Management is formal and hierarchical, but many instances of companies in operation throughout the European Union have a less rigid and more egalitarian structure, in ways that truly resemble organizations presently operating in the United States (Hopkins, 1997).

*Ethics in Latin America*

Individuals from Latin cultures reside in a variety of countries, including Mexico, South America, Central America, and the Caribbean. Collectively, their common Spanish ancestry bonds them. The predominant religion in these nations is Roman Catholic, which plays a significant role in their similar and often overlapping views regarding ethics and values. Family is a very high priority in Latin countries, and family obligations often supersede business responsibilities. Two vehicles that are essential to social mobility in Latin America are marriage and education (Hopkins, 1997). Also of great significance in Latin cultures is one's social position, which effectively determines the extent to which others in your community respect you. As in other regions of the world, education, well-mannered behavior, and land ownership are indicators of a person's social status.

The value system in Latin American companies typically emphasizes status and rank. Loyalty is expected, and there is great respect for managerial authority. Gomez (1993) explained that participative management style is not prevalent because that type of behavior makes Latin workers feel somewhat uncomfortable. Formality is important, but time and perspectives on work are quite relaxed in most Latin organizations. The practice of handing out gratuities or bribes for getting something accomplished is considered legal; these behaviors are an expected way of doing business in Latin firms (Hopkins, 1997). The custom of nepotism is common in Latin American companies, and it is not considered unethical behavior. It further illustrates the prominence of family in Latin cultures.

Mexico is one example of a Latin American country in which efforts have been made to transform its governmental and corporate institutions. In recent years, Mexico's economy has opened to free trade, subscribing to world organizations and agreements, and further encouraging direct foreign investment and private sector modernization (Organization for Economic Cooperation and Development [OECD], 1995). The government has reduced its size via the sale and auction of public corporations, thus promoting more activity in the private sector. Major political changes have occurred, electoral procedures are more transparent, and Mexico's democratic life has been strengthened. These changes have had an impact on Mexican society, its values and principles, and ethics within the public sectors (OECD). In Mexico, the stated moral principles of the public office are legality, honesty, loyalty, impartiality, and efficiency. This manifests a code of conduct that all public servants must observe. In the private sector, companies have also been paying more attention to developing and maintaining ethical standards among their workers. Attention to ethics in Mexico is currently viewed as critical to success in business (Adler, 2002).

## ETHICS AND TECHNOLOGICAL DEVELOPMENT IN OUR WORLD

Throughout this chapter, we have examined a wide array of philosophical beliefs, religious convictions, and expected social behaviors. Each of these has been discussed within the context of culture and ethics. In this final section of Chapter 2, we discuss the impact ethics and ethical behavior have on the development of a technological world in which we can all live. The authors of Chapter 3 introduce an array of more specific examples of ethical behaviors as related to the design and development of various technologies (for example, medical, agricultural, and transportation).

The relationship between society and its scientific and technological establishments is both simple and complex. Bird (2002) noted that members of the scientific and technical communities are part of society and are both honored and granted the privilege of pursuing their professional interests in large measure because of their contribution to the good of society. A major responsibility of professionals in science and technology is to ensure the accuracy and reliability of the information they develop, especially with regard to knowledge provided to persons in leadership/

decision-making roles (Bird). We have looked at what are considered to be ethical and moral behaviors throughout this chapter. Examples of unethical behavior are discussed in other chapters of this yearbook, but it is essential to review several of them here as they relate to scientific and technological development.

Contemporary science and technology have created the amenities of modern civilization. In countless ways, each has increased our capacity to control environmental forces and has given us visions of an even more prosperous future. Undeniably, science and technology have also increased both uncertainties about the future and our dependence on the numerous inventions and innovations produced. Stated differently, although we may have increased our ability to understand, predict, and control the natural environment around us, we may have actually lost the capacity to control the technologies we have created to help us out along the way. Swearengen and Woodhouse (2001) surmised the following:

> When negative consequences (of technological innovations) are immediate, stakeholders sometimes can assess costs and negotiate remedies and compensations—although when the costs and benefits accrue to different communities and demographic (ethnic) groups, both analysis and remediation can be difficult. When negative impacts of technological change manifest only after lengthy delay during which the offending technology is thoroughly adopted into commerce, correction becomes much more difficult. (p. 15)

Technology and its impact on society at large has been a prevalent theme in social, economic, and political thought for decades. Numerous essayists have debated about the extent to which continued technological growth will either enhance or hinder the survival of the human race. A diverse collection of writers has drawn our attention to the increasing complexity and rate of technological change in contemporary society. Almost habitually, they seem to be making projections filled with varying degrees of alarm. Their publications present the specter of an increasingly uncontrollable technology whose consequences cannot always be assessed or accurately predicted. We are drawn to infer that the human race is on a collision course to ruin and imminent destruction. Members of society must ethically examine the process of technological change and take actions to curtail its destructive and perhaps unintended consequences.

We are living amidst an array of technologies some persons contend are simply accidents waiting to happen. When the dark side of technology reveals itself, when disaster hurts innocent bystanders, and when technology causes more problems than it solves, society often points the finger of blame at its creators. When "accidents" do occur, the general public shakes its head while wondering what is happening to its "technological fix." Regardless, there is little chance that the industrialized nations of the world will turn their backs on the promises aligned with new technologies. We have all chosen a lifestyle sustained by high technology, but that doesn't mean we are not apprehensive about it (Markert and Backer, 2003).

Most of us place our faith and trust in the scientific and technological experts, and we assume that their research practices are ethical. Science and technology are not isolated; they are embedded in the context of social values, human interests, and political objectives. As such, they are subject to public scrutiny using ethical and social norms. Several perspectives on the types of actions we should regard as improper behavior in the arena of science and technology laboratories are presented in Drenth's (2002) work. He distinguished the following four subcategories of behaviors considered to be unethical in the conduct of scientific and technological research:

1. Unethical behavior including fraud (fabrication and falsification of data), deceit (deliberate use of improper sampling techniques), and infringement of intellectual property rights (plagiarism).
2. Improper or imprudent behavior vis-à-vis subjects, including not taking full account of the requirement of informed consent, open or hidden discrimination, and negligence of duty to exercise care in animal research.
3. Careless behavior with respect to the general public and the media, such as too optimistic or unjustified popular reports and interviews, negligence in cases of misquotations by the press, and taking no action in case of wrong or biased interpretations by colleagues or in the media.
4. Disregard of good practice rules such as justified authorship, proper sequence of authors on a published (or working) document, proper citations, correct dealing with secrecy or delay of publication in the interest of the research sponsor, and avoiding conflicts of interest. (pp. 8–9)

Indeed, these examples may seem overly simplistic, but the high stakes consequences of this type of misconduct in a world in which the pace of technological innovation is so accelerated are beyond dire. In many ways, the extent to which human beings around the world in all cultures are able to experience their lives as meaningful and satisfying is directly related to the practice of ethical and moral conduct across the entire landscape of science, technology, and engineering disciplines. Drenth (2002) defended the need for more international scientific and technological collaboration on the grounds of moral (and ethical) obligations of the Western and economically advanced countries to support and strengthen research and development capabilities in economically less developed nations. In the long run, aid and collaborative partnerships, focused on ethics and research themes that have an international character (for example, environment, health, infectious diseases, trade, transportation, security), will likely become the best precondition for peaceful coexistence in our world.

The most critical ethical challenges of the twenty-first century will undoubtedly center on the importance of intangible and perhaps obscure factors in shaping a technological world and civilization worth living in. Persons living in all corners of the planet are experiencing heightened tension and intense personal anxiety. Many of us worry that the human race has lost its way and is on a collision course with doom. "What some observers perceive as the contemporary crisis of society, east and west, is said to be due not so much to material constraints, lack of necessary techniques, or lack of information, but rather to a shortage of virtue" (Swearengen and Woodhouse, 2001, p. 16). As we learned earlier in the discussion of the ethics and values revered by various cultural groups, a good and fulfilling life does not revolve solely around more possessions, or better technologies, or even more extensive knowledge. In the end, personal character, public harmony, social ethics, and moral qualities must be interwoven with our technologically dependent, economic well-being to sustain a venerable civilization.

To conclude this discussion, let us review the intriguing principle of *kenosis*, a Greek word for self-emptying, but a term that also has meaning in theology. South African cosmologist George F. R. Ellis proposes that:

> The foundational line of true ethical behavior, its main guiding principle valid across all times and cultures, is the degree of freedom from self-centeredness of thought and behavior, and willingness to freely give up one's own self-interest on behalf of others . . . there is an ethical underpinning to the universe as well as a physical one . . . a

benevolent Creator arranged things just so intelligent beings could experience kenosis. (cited in Gibbs, 1995, p. 55)

As we in the technology education profession continue to devise ways for our students to learn about the role of ethics in the development and use of technology in their world, we must move beyond a United States' perspective. Social concepts are complex and often vague, and ethical issues are challenging to deal with. We have learned that some of the salient themes for ethical rules include integrity, responsibility, fairness, honesty, caring, protection from harm, disclosure, and so forth. Our next accomplishment will feature an understanding of how cultures other than our own apply meaning and value to these themes in activities related to the process of scientific discovery and technological innovation.

## REFLECTION QUESTIONS

1. What role did ethics and values play in the founding of the United States?
2. What limits should be placed on the display of religious documents in government buildings in the United States?
3. How does the process of immigration to the United States in the twenty-first century compare to that of the nineteenth century? Should immigrants be encouraged to assimilate mainstream cultural values or to preserve their own cultural values?
4. How would you explain ethics in the United States to someone visiting from another country who inquired about the ethical behavior of government or corporate leaders?
5. If you were living in another country because of your job, how would you be influenced by the ethical standards of a different culture?
6. What ethical standards should be used when businesses owned by United States's citizens are operated in other lands? What if business practices that would be illegal in the United States are acceptable practice in this international environment?
7. How prevalent is the Protestant Work Ethic in the twenty-first century United States?

8. Would you expect there to be an ethic comparable to the Protestant Work Ethic in Asia, Africa, or Latin America? What might be the basis for this ethic?
9. What impact has the Protestant Work Ethic and other ethical belief systems had on the prosperity of the United States?

# REFERENCES

Adler, I. (2002). Walking the walk: Can teaching ethics increase business profits? *Business Mexico.* Retrieved September 30, 2002, from http://www.mexconnect/mex_/travel/bzm/bzmadler38.htm

Andrew, T. (1988, May 16). Samsung: South Korea marches to its own drummer. *Forbes,* 84-89.

Bird, S. (2002). Science and technology for the good of society? *Science and Engineering Ethics, 8*(1), 3-4.

Bunch, L. G., Crew, S. R., Hirsch, M. G., & Rubenstein, H. R. (2000). The American presidency: A glorious burden. Washington, DC: Smithsonian Institution Press.

Byron, S. J. (1977, November). The meaning of business ethics. *Business Horizons,* 32.

Cherry, M., & Matsumura, M. (1998). Ten myths about secular humanism. *Free Inquiry Magazine, 18*(1). Retrieved September 30, 2002, from http://www.secularhumanism.org/library/fi/cherry_18_1.01.html

Commission reviews Middle Eastern perceptions of the United States. (2002, May 24). U.S. Department of State. Retrieved September 30, 2002, from http://www.state.gov/r/adcompd/11844pf.htm

Council for Secular Humanism. (2002). *What is secular humanism?* Retrieved October 5, 2002, from http://www.secularhumanism.org/intro/what.html

Cua, A. S. (1992). Confucian ethics. In L. C. Becker & C. B. Becker (Eds.), *Encyclopedia of ethics.* New York: Garland.

Drenth, P. (2002). International science and fair-play practices. *Science and Engineering Ethics, 8*(1), 5-11.

Enderle, G. (2000). Ethical guidelines for the reform of state-owned enterprises in China. In O. E. Williams (Ed.), *Global codes of conduct: An idea whose time has come.* Notre Dame, IN: University of Notre Dame Press.

Furnham, A. (1990). *The protestant work ethic: The psychology of work-related beliefs and behaviours.* London: Routledge.

Gibbs, W. W. (1995, October). Profile: George F. R. Ellis. Thinking globally, acting universally. *Scientific American, 273,* 50, 54-55.

Gluckman, M. (1965). *Politics, law and ritual in tribal society.* Chicago: Aldine.
Gomez, J. E. A. (1993). Mexican corporate culture. *Business Mexico, 3*(8), 8-9.
Goodman, N. (1990). *Doing business in Japan.* Randolph, NJ: Global Dynamics.
Gudkov, L. (2002, February). How are we any worse? Analysis of the roots of anti-Americanism in Russia. *Russia Weekly.* Retrieved October 6, 2002, from http://www.cdi.org/russia/193-7.cfm
Hopkins, W. E. (1997). *Ethical dimensions of diversity.* Thousand Oaks, CA: SAGE.
Hussain, M. (2001, October 19). Anti-Americanism has its roots in U.S. foreign policy. *Inter Press Service.* Retrieved October 6, 2002, from http://www.commondreams.org/views01/1019-05.htm
Kurtz, P. (2002). Secular humanism: A new approach. *Free Inquiry Magazine, 22*(4). Retrieved October 5, 2002, from http://www.secularhumanism.org/library/fi/kurtz_22_4.htm
Maccoby, M., & Terzi, R. (1979). What happened to the work ethic? In W. Hoffman and T. Wyly (Eds.), *The work ethic in business.* Cambridge, MA: O, G & H.
Magesa, L. (1997). *African religion: The moral tradition of abundant life.* Maryknoll, NY: Orbis.
Markert, L. R., & Backer, P. R. (2003). *Contemporary technology: Innovations, issues and perspectives.* Tinley Park, IL: Goodheart Willcox.
McElroy, J. H. (1999). *American beliefs: What keeps a big country and a diverse people united.* Chicago: Ivan R. Dee.
McLaughlin, R. (1990). *Marketing in the United Kingdom.* (Overseas Business Reports, No. 5), Washington, DC: U.S. Department of Commerce, International Trade Administration.
Nash, R. J. (2002). *Real world ethics: Frameworks for educators and human service professionals.* New York: Teachers College Press.
National Science Board. (2000). *Science & engineering indicators–2000.* Arlington, VA: National Science Foundation.
Oates, W. (1971). *Confessions of a workaholic: The facts about work addiction.* New York: World Publishing.
Organization for Economic Cooperation and Development (OECD). (1995). Ethics and corruption. *The management of ethics and conduct in the public service: Mexico.* Retrieved September 30, 2002, from http://www1.oecd.org/puma/ethics/pubs/ethics.mx.htm

Padover, S. K. (1960). *The world of the founding fathers: Their basic ideas on freedom and self-government.* New York: Thomas Yoseloff.

Rokeach, M. (1968). *Beliefs, attitudes and values.* San Francisco: Jossey-Bass.

Singh, J. P., & Hofstede, G. (1990). Managerial culture and work-related values in India. *Organization Studies, 11*(1), 75-106.

Swearengen, J. C., & Woodhouse, E. J. (2001, Spring). Cultural risks of technological innovation: The case of school violence. *IEEE Technology and Society Magazine, 20*(1), 15-28.

Swenson, W. (2000). Raising the ethics bar in a shrinking world. In O. E. Williams (Ed.), *Global codes of conduct: An idea whose time has come.* Notre Dame, IN: University of Notre Dame Press.

Turner, V. W. (1966). Ritual aspects of conflict in African micropolitics. In M. J. Swartz (Ed.), *Political anthropology.* Chicago: Aldine.

U.S. Commission on Immigration Reform (USCIR). (1997). *Becoming an American: Immigration and immigration policy.* Report to Congress. Retrieved October 4, 2002, from http://www.utexas.edu/lbj/uscir/becoming/intro.html

United Nations Department of Economics and Social Affairs Division for Public Economics and Public Administration (UN/DESA/DPEPA). (2001). *Public service ethics in Africa.* New York: United Nations.

Weber, M. (1904, 1905). *The protestant ethic and the spirit of capitalism.* Translated by T. Parsons. New York: Charles Scribner's Sons.

Wolfe, A. (2001). *Moral freedom: The search for virtue in a world of choice.* New York: W. W. Norton.

*Words of our American founding fathers.* (2002). Retrieved September 30, 2002, from http://www.stephenjaygould.org/ctrl/quotes_founders.html

# Ethics and the Design and Development of Technological Systems

## Chapter 3

Michael A. DeMiranda
Colorado State University
Fort Collins, CO

Jack Wescott
Ball State University
Muncie, IN

Len S. Litowitz
Millersville University
Millersville, PA

Myra N. Womble
The University of Georgia
Athens, GA

Mark Sanders
Virginia Tech
Blacksburg, VA

Stephanie Williams
The University of Georgia
Athens, GA

Richard D. Seymour
Ball State University
Muncie, IN

---

The *Standards for Technological Literacy: Content for the Study of Technology* (International Technology Education Association [ITEA], 2000) clearly supports the importance of ethics as a facet of technological literacy. Instructional content related to ethics should not, however, be isolated from the context of a technological world. Teachers do not typically have time to include new units of instruction in their classes and even if they did, without clear connections to practical applications, discussion of ethics in technology education classes would likely have little impact.

Case studies or vignettes that highlight the role of ethics and character in the design and development of technological systems are a valuable instructional resource. They can be used to describe the significance of ethics in a technological workplace and highlight the dilemmas and problems that can occur when ethics are absent. Classroom teachers and teacher educators can use them in a variety of ways to introduce the topic of ethics as technical content is presented.

Included in this chapter are case studies related to each of the seven areas of the designed world described in the standards document.

Materials are also provided to explain the background and context for the scenarios, and suggested questions for discussion follow each section.

## MEDICAL TECHNOLOGIES
*Michael A. DeMiranda*

The infusion of technology into the practice of medicine during the past 50 years has created a new medical era. Advances in material science technology have led to the production of artificial limbs, heart valves, and blood vessels, thereby permitting advances like "spare-parts" surgery. Numerous patient disorders are now routinely diagnosed using a wide range of highly sophisticated imaging technology, and the lives of many patients are being extended through significant improvements in resuscitative and support technology, such as respirators, pacemakers, and artificial kidneys.

These technological advances, however, have not been benign. They have had significant moral consequences and have created numerous ethical dilemmas. Advances in medical technologies have provided healthcare providers with the ability to use cardiovascular assistive devices, perform organ transplants, and maintain the breathing and heartbeat of terminally ill patients. In doing so, society has been forced to reexamine the meaning of such terms as quality of life, heroic efforts, informed consent, and acts of mercy (Benner, 2003). In addition, considerations such as moral issues, the rights of patients to refuse treatment (living wills), and participation in experimental treatments (informed consent) have complicated the ethical landscape. Technological advances have made the moral dimension of health care and indeed the study, design, and development of medical technologies more complex (Reiser, Dyck, and Curran, 1997).

The purpose of this chapter contribution is to examine some of the moral and ethical tensions that arise when humans design, perfect, and implement medical technologies. The objective, however, is not to provide solutions or recommendations that will resolve all of these tensions, but rather to demonstrate that each technological advance has consequences that affect the very core of human values. The dilemmas and issues faced in the use of medical technologies are described in a simple case study following the framing of issues related to medical technologies, advances, and ethics.

# MEDICAL ADVANCES, TECHNOLOGY, AND ETHICS

Medical advances, technology, and ethics are interrelated concepts in the world of human innovation and accomplishment. Technology can be characterized as a human achievement of exemplary innovation and utility that is often quite distant from the human accomplishment of ethical values (Hoffman, 2002). The intersection of advances in medical technology and ethical values was accelerated in the twentieth century by the advances in basic science and the intense cross-fertilization of scientific and technological discoveries, such as electrical measurement techniques, sensor development, nuclear medicine, and diagnostic ultrasound (Bennett, 1977). For example, in 1903, William Einthoven devised the first electrocardiograph and measured the electrical changes that occurred during the beating of the heart. In the process, Einthoven initiated a new age for both cardiovascular medicine and electrical measurement techniques (Enderle, Blanchard, and Bronzino, 2000; Snellen and Hollman, 1996).

New medical technologies followed, but none was more significant than the scientific knowledge of X rays and the development of a technological device to generate them. When W. K. Roentgen discovered his "new kind of rays," the human body was opened to medical inspection (Aronson, 2000). The power this technological innovation gave physicians was enormous. The X ray permitted them to diagnose a wide variety of diseases and injuries accurately. In addition, being located within the hospital, it helped trigger the transformation of the hospital from a historically passive receptacle for the sick to an active curative institution for all the citizens of American society (Bronzino, Smith, and Wade, 1990; Howell, 1996).

As medical technology in the United States blossomed, so did the prestige of American medicine. From 1900 to 1929, Nobel Prize winners in physiology or medicine came from Europe. In the period from 1930 to 1939 prior to World War II, seven Americans were honored with a Nobel Prize. During the postwar period from 1945 to 1975, thirty-nine American life-scientists were honored, and from 1975 to 1995, thirty-one more were honored. The accomplishments recognized by these awards were made possible by the advanced medical technology available to these clinical scientists (Levinovitz and Ringertz, 2001).

Rapidly expanding medical technologies available to the medical profession also advanced the development of complex surgical procedures.

The Drinker respirator was introduced in 1927 and the first heart-lung bypass machine in 1939. In the 1940s, the cardiac catheterization and angiography procedures were made possible through advances in material science. The use of a cannula threaded through an arm vein and into the heart with the injection of radiopaque dye for X-ray visualization made seeing the heart and lung vessels and valves possible (Cruse, 1999; Howell, 1996).

The list of technological advances in medicine goes on and on. Medical applications of the electron microscope, body scanners to detect tumors and other abnormalities, and computer integration with tomography and magnetic resonance imaging have all enhanced the capabilities of medical professionals to identify and treat diseases and injuries. When body parts are damaged beyond repair, prosthetic devices such as artificial heart valves, artificial blood vessels, functional electromechanical limbs, and reconstructive skeletal joints can be used as replacements (Lalan, Pomerantseva, and Vacanti, 2001).

Medical technology improvements in the past 50 years have exceeded advances during the previous 2000 years. In a culture steeped in science, it appears the trend in medical technologies and their advance and impact will continue. However, the social and economic consequences of this vast outpouring of information and innovation must be fully understood if medical technologies are to be implemented effectively and efficiently (Mauzur, 2002).

## MEDICAL TECHNOLOGY AND ETHICS

Innovations in medical technology have endowed modern medical care providers with the ability to sustain and prolong life, repair damaged body parts, peer into the human body, cure and treat many diseases, and otherwise ameliorate a wide range of undesirable physical and mental conditions for which little could have been done in the recent past. However, to see only the benefits of medical technologies is to fail to see the whole picture. Along with these impressive advances and technological innovations come a number of difficult and often perplexing ethical issues (Beach and Morrison, 2002; Blake, 1988; Rajput and Bekes, 2002).

In the practice of medicine, moral dilemmas are not new. They have been present throughout medical history. As a result, over the years there have been efforts to provide a set of guidelines for those responsible for patient care. These efforts have resulted in the development of specific codes of professional conduct. On behalf of all professionals who use medical technologies, the World Medical Association adopted a version of the Hippocratic Oath titled the Geneva Convention Code of Medical Ethics in 1949 (Rosenblatt, 2000). In 1999, the House of Delegates of the American Medical Association revised a set of Principles of Medical Ethics (World Medical Association, 1999). These codes take as their guiding principle the concepts of service to humankind, respect for human life, and prevention of harm and malice.

Although the established codes of conduct are useful in promoting the ethical treatment of patients and the conduct of the health service providers, the codes fail to provide answers to some of the difficult moral dilemmas involving the use and appropriation of medical technologies (Bronzino, 1992). For example, all of the fundamental responsibilities of a physician cannot be met at the same time. When a patient suffering from massive trauma to the brain is kept alive by means of artificial life-support equipment and this equipment is needed elsewhere for the recovery of a surgical patient, it is not clear in the code of ethics how such decisions should be resolved. Ethical dilemmas of the right-versus-right type described in Chapter 1 frequently arise within the context of modern medical practice.

## LIFE AND DEATH DECISIONS

A fitting place to examine the way ethics and the moral aspects of medical practice intersect with the use of advanced technologies is to scrutinize how modern life-support systems found in intensive care units of hospitals impact medical decisions. Consider the case of a patient who sustained a serious head injury in an automobile accident. When the emergency medical personnel and ambulance arrived on the accident scene, the patient was unconscious but still alive with a beating heart. After the victim was rushed to the emergency ward of a local hospital, the resident in charge verified the stability of the patient's vital signs of heartbeat and respiration during the initial examination. The physician ordered a computerized tomography scan to indicate the extent of the head injury. The results of this technological procedure clearly showed extensive brain damage. When the electroencephalogram (EEG) was obtained from the scalp electrodes placed around the patient's head, it was noted to be significantly abnormal. In a case such as this, obvious questions arise: What is the status of the patient? Is the patient still alive?

Alternatively, consider the events encountered during an open-heart surgery. During this procedure, the patient was placed on a heart-lung bypass machine while the surgeon attempted to repair a damaged heart valve. As time passed on this long and complex procedure, the EEG monitor sounded a loud alarm that alerted the operating room staff that the normal pattern of electrical activity displayed in the brain at the beginning of the operation had suddenly changed to a straight line indicating weak or no electrical activity. However, because the heart-lung bypass machine was maintaining the patient's vital signs, viability of body tissues was being maintained. Once again, questions present themselves regarding what the surgeon and her staff of medical technicians should do. Is the patient alive or is the patient dead?

Circumstances in which medical technologies can sustain or support bodily functions have required that medical professionals and indeed society reexamine the definition of death. In essence, advances in medical technology that delay or prevent death have actually hastened its redefinition (Penticuff, 1990).

The definition of death has historically been closely related to medical science. For the centuries prior to artificial respirators, death was defined as the absence of breathing. It was often believed that the spirit of human existence resided in the spiritus (breath); its absence was indicative of death. With advances in knowledge of human physiology and the development of medical technologies and techniques to revive a person who was not breathing, attention then turned to the pulsating heart as the focal point in determination of death. However, this view changed with the development of medical devices and technological advances in supportive therapy, resuscitation, cardiovascular assistive devices, and organ transplantation (Dickerson, 2002).

As the knowledge of the human organism increased, it became apparent that blood circulation, and particularly the delivery of oxygen provided by blood flow, was critical for support of life. The advent of diagnostic technology to monitor patient blood gas levels led to the understanding that all vital organs require oxygen, hence any organ deprived of oxygen for periods greater than three minutes would begin to suffer damage. The critical functions of the brain are particularly sensitive to oxygen deprivation and irreversible damage to brain tissue results when blood circulation is interrupted. Consequently, the evidence of death began to shift from the pulsating heart to the vital functioning of the brain. After the technology of the EEG was introduced to monitor the brain's activity, another factor was introduced to the definition of death. The moral resolution associating death with lack of brain activity was based on reasoning that when the brain is irreversibly damaged, so are the functions that are identified with self and being human, such as memory, feeling, thinking, and knowledge. As a result, it became widely accepted that the termination of activity of the lungs, heart, and brain should define clinical death. The irreversible cessation of functioning of these three major organs became the standard for anyone to be pronounced dead (Humber, 1991; Ott, 1995).

With the development of advanced medical technologies that supported artificial respiration, the medical profession encountered an increasing number of situations in which a patient with irreversible brain damage could be maintained almost indefinitely. Once again, new medical technology advancement created the need to reexamine the definition of death.

## THE CASE OF KAREN ANN QUINLAN

In 1975, Karen Ann Quinlan suffered severe brain damage that resulted in her entering a chronic vegetative state in which she had little cognitive function. Upon consulting medical experts and specialists, her parents requested permission from the New Jersey courts to have the medical technologies that sustained her respiratory functions disconnected. The courts of New Jersey upheld the parents' request that the respirator sustaining their daughter's life could be removed and disconnected. However, the nurse in charge of her care in the Catholic hospital opposed the decision and in anticipation began to wean Karen from the respirator so that by the time the court's decision was handed down and the respirator disconnected, Karen could remain alive without its assistance. Karen Ann Quinlan did not die. She remained alive for 10 years and in 1985 she died of acute pneumonia. Antibiotics, which would have fought the pneumonia, were not given (Battelle, 1976; Casey, 1976; Kohl, 1976; Washington Post, 1985).

The following questions could be considered with regard to the Karen Quinlan case:

1. When did Karen Quinlan really die?
2. Critique the parents' decision to remove their daughter from the respirator.

3. Critique the hospital nurse's decision to wean Karen from the respirator.
4. How would you decide what to do with regard to keeping a close family member on life support?
5. If you were in Karen Quinlan's situation, what action would you want taken?

The Karen Quinlan case illuminates the elusive resolution of human dilemmas that result from the capabilities of modern medical technologies. Is an individual in a state of neocortical death but with a functioning respiratory and circulatory system one whose respiration and circulation are mechanically maintained? It is a matter that society must decide. The ability to cope with complex ethical issues is clearly needed to resolve the dilemmas presented by advanced medical technologies.

## FUTURE OF MEDICAL TECHNOLOGY

The field of medical technology encompasses some of the most advanced applications of human ingenuity, design, and innovation ever assembled through the application of modern science; however, the technologies of acute care medicine receive the most discussion and generate the most ethical dilemmas. These high-profile technologies include the supportive and resuscitative devices designed to minimize levels of morbidity and mortality.

The rescue natures of the high-profile medical technologies are in sharp contrast to another class of lesser-known medical technologies associated with preventive medicine. Rescue medical technologies include all types of medical technologies that attempt to restore health and normal functioning to individuals after illness or disability has occurred. Preventive medical technologies are based on a completely different premise. Helwege (1996) defined preventive technologies as attempting to prevent or significantly delay the occurrence of illness and disability in the first place. The goal of preventive medical technologies is to make rescue medical technologies unnecessary.

Although the attainment of the ideal will never be entirely met, the extent to which the ideal is approximated will depend on levels of adoption of preventive technologies rather than the technology of rescue. The significance of the moral and ethical issues associated with preventive medicine have not been fully recognized (Callahan, 2002). The popular image of preventive medicine falsely depicts it as a largely nontechnological form of medical care. Granted, the technologies of preventive medicine are likely to be substantially different than those of acute care rescue medicine; however, some of the most sophisticated medical technologies will play a role in the future of preventive medicine, and the related ethical issues will be equally challenging as those posed in the practice of rescue medicine.

## THE CASE OF NANOMEDICAL TECHNOLOGY

A small, palm-sized computer that could recognize handwriting was once considered an advanced technical achievement, but that's nothing compared with a handheld that recognizes something more personal and significant—DNA. In an overview of the National Nanotechnology Initiative, Meyyappan (2000) described innovations developed by researchers at Northwestern University's Institute for Nanotechnology Center. The handheld DNA Detector can spot the DNA of nasty diseases in a matter of minutes instead of days. The device can be thought of as a Personal Digital Assistant that provides all of the detection information that would be gathered over the course of an ordinary doctor's office visit.

The diminutive device holds great promise for the delivery of medical care and for the detection of biological warfare agents. It not only allows doctors to make on-the-spot diagnoses during patients' office visits, but it also costs a fraction of conventional diagnostic testing, which requires a doctor to deliver samples of blood to a laboratory for polymerase chain reaction testing. Polymerase chain reaction is a Nobel Prize-winning invention, but it has drawbacks. The test equipment is expensive, and the process is painstakingly time-consuming. While the diagnostic equipment is analyzing a sample for days, running and rerunning tests, patients are stressed as they await critical results. Furthermore, the technology is not mobile, so a healthcare worker cannot take one into the field. Ultimately, a DNA handheld could rapidly detect virtually any viral, bacterial, or genetic agent with a known DNA sequence, including biological weapons and genetic markers for cancer (Williams, 1999; Zajtchuk, 1999).

Consider a scenario in which DNA handhelds became a standard part of the application process for employment. Companies could quickly screen applicants to determine probability of absences from work due to health issues. Healthcare costs could also be minimized, particularly for large companies with self-insurance systems. Because health information from DNA handhelds would be in the form of digital data, it could easily be transmitted or maintained in electronic databases. After a health concern was identified for an individual, it might follow that person throughout his or her life, impacting career opportunities and influencing everything from social status to eligibility to obtain credit.

1. Comment on whether the use of DNA handhelds should be regulated.
2. If you were the owner of a business or corporation, how would you respond to a request by your personnel manager to use DNA handheld devices to screen prospective employees?
3. Describe the major issues related to the establishment of electronic databases that would store DNA information on individuals. Construct an argument in favor of or against the development of these information systems.

## THE CASE OF MICRO ELECTROMECHANICAL SYSTEMS (MEMS)

Once designed for sensor application in modern computer-controlled automobiles, the application of MEMS to biomedical technologies has never been brighter. An example of the application of MEMS technology to preventive medical applications is implantable patient monitoring systems. The implantable device is made of a highly sophisticated fiber-optic pressure sensor made by Fiso Technologies. The device employs a MEMS-based sensing catheter that provides in vivo physiological body function measurements (Polla et al., 2000). On the immediate horizon of this technology is an implantable MEMS lab-on-a-chip device that will permit healthcare providers to perform point-of-care diagnostics on patients without the time and cost of current methods.

At the Cleveland Clinic Foundation in Ohio, a group of engineers and technologists are working on MEMS sensors and antennas that will let neurosurgeons accurately study and control the human spine. This MEMS-based system will be attached across four vertebras. The MEMS will contain a strain gauge on the vertebra's outside and a pressure sensor/actuator to monitor bone fusion and increase it through stimulation. The strain gauge, pressure sensor, and actuator stimulator will all be connected to a tiny radio transmitter to communicate with monitoring equipment outside the person's body. MEMS technology is being harnessed to realize a host of implantable mechanisms that can stimulate paralyzed limbs, improve the treatment of diseases like epilepsy and Parkinson's, diagnose bacterial and viral agents, determine the safety and efficacy of drugs administered, and speed up drug delivery to the point of direct need in the body (Kotzar et al., 2002).

Indeed, in the world of medical engineering design and technology, we are witnessing just the tip of the iceberg of what lies ahead in medical technologies. Specifically, the MEMS technology when applied to medical applications has shown promising results because of its unique and versatile combination of mechanical, electrical, and physical properties. Consider, for instance, a scenario in which children were implanted with MEMS devices at birth. These devices could be used to monitor health throughout infancy, childhood, and adolescent years. If left in place, they could also assist in diagnosing adult diseases. With minor software modifications, these implanted chips could also be used to transmit location and deter kidnapping and other crimes against children. The MEMS devices implanted in adults might even be programmed to interact with proximity sensors to unlock doors in buildings, control entry and ignition on automobiles, and facilitate financial transactions by replacing credit and debit cards.

1. What are the advantages and disadvantages of implanting MEMS devices in children?
2. MEMS devices could be developed that are tiny enough to be inserted with a simple injection. Should this happen, describe your reaction to requirements for all infants to have MEMS devices implanted prior to leaving the hospital after birth.
3. If implanting of MEMS devices became commonplace, who should manage the system? A federal agency? State agency? Local government? Private companies?

## SUMMARY

In this section, we have examined some of the many ethical issues posed by recent innovations in medical technology. Unfortunately, along with the increase in ability to benefit humans, the advances in medical technology have made it more difficult than ever before in the history of medicine to know just what constitutes being beneficent or maleficent. This situation was illustrated in the cases presented, and the magnitude of the ethical questions facing healthcare professionals and society as a whole will only get larger.

*Ethics and the Design and Development of Technological Systems*

If a technologically literate society is defined as being able to use, manage, and apply technology, the definition becomes increasingly complex when dealing with medical technologies. The challenges are compounded by the ethical dilemmas related to redefining death, deciding when the appropriate application of a medical technology is justified based on a person's prognosis for a quality life, or meeting the growing demand for transplantable human organs. The values and ethical standards we as a society adopt are keenly dependent on our knowledge of the technologies that intersect our lives. Furthermore, not everyone can be provided with everything they need or want by way of medical care, and this is especially true of some of the highly sophisticated medical technologies that have recently become available. Society must determine which medical resources will be available to all and which privileges will be allowed to those who have private wealth substantial enough to purchase whatever they desire.

The ethical issues raised in this section should not be taken to imply that medical technology is wholly a bane or wholly a boon. Medical technology does have the capacity to provide our society with massive benefits, but utilizing that capacity poses hard ethical questions. Who is to decide when, where, and for how long life is to be prolonged? What will be measures for which allocation of leading medical technologies are appropriated across all segments of our population? When does the technology begin to overshadow the limit of basic human dignity? If our society wants health care that is steeped in the technological innovations of our time, that is both efficient and humane, then it cannot avoid the responsibility of tackling these ethical questions. Technology education should strive to prepare students to engage in this debate.

# BIOTECHNOLOGIES

*Michael A. DeMiranda*

The *Chambers Science and Technology Dictionary* defines biotechnology as the use of organisms or their components in industrial or commercial processes, which can be aided by the techniques of genetic manipulation in developing, for example, novel plants for agriculture or industry (Walker, 1988). Despite the inclusiveness of this definition, the biotechnology sector is still often seen as largely medical or pharmaceutical in nature, particularly among the general public. Although to some extent the huge research budgets of the drug companies and the widespread familiarity of their products makes this understandable, it does distort the full picture (U.S. Congress, Office of Technology Assessment, 1987).

The inclusion of biotechnologies as a technical content area in technology education is relatively new. This being the case, this section provides an introduction to biotechnology for technology educators and discusses the ethical and often moral appropriation, application, and privilege humans have created in manipulating the natural world of our existence. Case studies involving ethical issues are organized around the four main categories of biotechnologies: (*a*) agricultural biotechnology, (*b*) pharmaceutical biotechnology, (*c*) environmental biotechnology, and (*d*) industrial biotechnology.

## EARLY HISTORY

Biotechnology can be viewed as a group of useful, enabling technologies with wide and diverse applications in industry, commerce, and the environment. Historically, biotechnology can be traced to when early humans became domesticated enough to breed plants and animals; gather and process herbs for medicine; make bread and wine and beer; create many fermented food products, including yogurt, cheese, and various soy products; create septic systems to deal with their digestive and excretory waste products; and create vaccines to immunize themselves against diseases. Archeologists keep discovering earlier examples of each of these examples of biotechnology, but most of these processes go back to the years 5000 to 10,000 B.C.

In the artisan applications of biotechnology, the techniques of manufacture were well-known and understood, but the molecular mechanisms went unknown in the past. In recent times, the progress and advances of microbiology and biochemistry have become better understood and, as a result, processes have become more controlled and improved. The improvement in basic understanding has led to the development of many new biotechnology products that are now commonplace in our lives. Although the history of experimentation and use of biological technologies to meet human needs is long, the abbreviated timeline in Table 3-1 illustrates that biotechnology has been a part of human existence for many years (Fiechter, Beppu, and Beyeler, 2000).

## Ethics and the Design and Development of Technological Systems

Table 3-1. Timeline of Significant Events in the History of Biotechnology

| Date | Event |
|---|---|
| 1750 B.C. | The Sumerians brew beer. |
| 500 B.C. | The Chinese use moldy soybean curds as an antibiotic to treat boils. |
| 100 A.D. | Powdered chrysanthemum is used in China as an insecticide. |
| 1590 | Janssen invents the microscope. |
| 1663 | Hooke first describes cells. |
| 1675 | Leeuwenhoek discovers bacteria. |
| 1797 | Jenner inoculates a child with a viral vaccine to protect him from smallpox. |
| 1830 | Proteins are discovered. |
| 1833 | The first enzymes are isolated. |
| 1855 | The Escherichia coli (E. Coli) bacterium is discovered. It later becomes a major research, development, and production tool for biotechnology. |
| 1869 | Miescher discovers DNA in the sperm of trout. |
| 1877 | Koch develops a technique for staining and identifying bacteria. |
| 1879 | Fleming discovers chromatin, the rodlike structures inside the cell nucleus that later came to be called chromosomes. |
| 1906 | The term "genetics" is introduced. |
| 1919 | A Hungarian agricultural engineer first uses the word "biotechnology." |
| 1928 | Fleming discovers penicillin, the first antibiotic. |
| 1940 | American Oswald Avery demonstrates that DNA is the "transforming factor" and is the material of genes. |
| 1946 | Discovery that genetic material from different viruses can be combined to form a new type of virus, an example of genetic recombination. |
| 1953 | *Nature* publishes James Watson's and Francis Crick's manuscript describing the double helical structure of DNA, which marks the beginning of the modern era of genetics. |

*continued*

*continued from previous page*

| Date | Event |
|------|-------|
| 1956 | Kornberg discovers the enzyme DNA polymerase I, leading to an understanding of how DNA is replicated. |
| 1964 | The International Rice Research Institute in the Philippines starts the Green Revolution with new strains of rice that double the yield of previous strains if given sufficient fertilizer. |
| 1970 | The first complete synthesis of a gene occurs. |
| 1976 | The tools of recombinant DNA are first applied to a human inherited disorder. |
| 1980 | The U.S. Supreme Court, in the landmark case *Diamond* v. *Chakrabarty*, approves the principle of patenting genetically engineered life forms, which allows the Exxon oil company to patent an oil-eating microorganism. |
| 1981 | A Chinese scientist is the first to clone a fish—a golden carp. |
| 1984 | The DNA fingerprinting technique is developed. |
| 1988 | Harvard molecular geneticists are awarded the first U.S. patent for a genetically altered animal—a transgenic mouse. |
| 1990 | The first federally approved gene therapy treatment is performed successfully on a 4-year-old girl suffering from an immune disorder. |
| 1993 | The FDA declares that genetically engineered foods are "not inherently dangerous" and do not require special regulation. |
| 1995 | The first baboon-to-human bone marrow transplant is performed on an AIDS patient. |
| 1998 | University of Hawaii scientists clone three generations of mice from nuclei of adult ovarian cumulus cells. |
| 2000 | Human genome is sequenced. |
| 2001 | An artificial heart is first implanted into a live human. Surgeons in Louisville, KY, report success of the first operation for a self-contained organ. |
| 2003 | The second and third generations of biotech crops expected to reach the market. |

(Historical timeline adapted from Fiechter, Beppu, and Beyeler, 2000, and information retrieved from http://www.biospace.com/articles/timeline_pre1900.cfm.)

## MODERN ERA OF BIOTECHNOLOGY

Since the middle of the twentieth century, advances in technology that enabled humans to see and manipulate the unseen world of cells, cell structure, and the fundamental building blocks of life have led to advances in the growth of biotechnology. The history between the beginnings of biotechnology and the advent of the twenty-first century included the isolation of DNA in 1869 by Friederich Miescher; the discovery of penicillin by Alexander Flemming in 1928; the discovery of the structure of DNA in 1953 by James Watson, Francis Crick, and Rosalind Franklin; the deciphering of the genetic code in 1961 by Marshall Nirenberg and H. Gobind Khorana; the first recombinant DNA experiments in 1973 by Walter Gilbert; the creation of the first hybridomas in 1975; the start of Genentech (the first biotechnology company) in 1976; the production of the first monoclonal antibodies for diagnostics in 1982; and the production of the first human therapeutic protein (humulin) in 1982 (Kristiansen and Ratledge 2001).

Why the rapid advances in the past 100 years? Rifkin (1998) argued that the main reason must be associated with the rapid advances in molecular biology and, in particular, recombinant DNA (rDNA) technology. This rDNA technology enabled technologists working in scientific laboratories to directly manipulate the heritable material of cells between different types of organisms creating new combinations of characteristics and functions not previously achievable by traditional breeding methods.

## PROFOUND DISCOVERIES AND ETHICAL DILEMMAS

Few technologies in the history of humankind have had the potential to impact the core foundations of life like biotechnologies. It is likely that applications of rDNA or genetic engineering will become the most revolutionary technology of our time (Macer 1990). The prospect of advances in biotechnology and genetic engineering, combined with the ability of medical technologies to detect hereditary defects in humans, a *priori*, suggests that gene therapy will someday be used to prevent predisposed diseases. Genetic engineering will also impact the development of biopharmaceutical drugs; vaccines for human and animal use; and the modification of microorganisms, plants, and farmed animals for improved and tailored food production.

Advocates of these biotechnology applications claim that in plant and animal breeding, these newfound technologies are much faster and have lower costs than traditional methods of selective breeding. Improvements to plant yields could result in a significant increase in our ability to feed a growing world population with reduced environmental impact. Higher yields allow less land dedicated to farming, thus reducing the use of pesticides and ground water depletion. These applications form, in many respects, the "acceptable" face of biotechnology, but elsewhere the science is all too frequently linked with unnatural interference and ethical questions (Bhardwaj, 1999; Macer, 2000).

Historically, biotechnology has primarily been used in production of food and medicine and to solve environmental problems. Modern biotechnology, based on rDNA technology, has a similar array of applications and can best be studied using the organizers of agricultural biotechnology, pharmaceutical biotechnology, environmental biotechnology, and industrial biotechnology. Ethical concerns and dilemmas arise in each of these areas and contribute to the complexity of citizenship in a technological world.

## ETHICAL ISSUES IN AGRICULTURAL BIOTECHNOLOGY

Applications of biotechnology in agriculture have attracted a fair amount of attention within society. Average citizens have become familiar with biodegradable products, industrial packaging using environmentally friendly materials such as cornstarch-based packing materials rather than polystyrene, and the use of recyclable or sustainable materials. However, biotechnology has also been a catalyst for consideration of bioethical issues, and the two words, biotechnology and bioethics, have become intertwined in the minds of many people (Macer, 1990).

Applications of genetic technology are actually quite commonplace in daily life. For example, crops such as chicory, maize, cotton, tomatoes, soya beans, squash, and potatoes have been modified or "engineered" to have specific traits. The information in Table 3-2 illustrates how some common crops have been modified to achieve desired traits and also lists the companies that own the rights or genetic patents for the food plants presented (Juma, 1989).

Table 3-2. Genetically Engineered Crops, Traits, and Commercial Interest

| Crop | Trait | Company |
|---|---|---|
| Tomato | Modified ripening | Zeneca Plant Science |
| Tomato | Delayed ripening | Calgene |
| Soya beans | High oleic acid content of oil | Du Pont |
| Cotton | Herbicide tolerance and insect resistance | Calgene |
| Potato | Insect resistance | Monsanto Company |
| Maize | Male sterile | Plant Genetic Systems |
| Oilseed rape | Male sterile/fertility restorer | Plant Genetic Systems |

A majority of the genetic engineering now being used commercially is in the agricultural sector. Plants are genetically engineered to be resistant to herbicides, to have built-in pest resistance, and to convert nitrogen directly from the soil. Related work is being pursued by entomologists who develop genetically engineered insects to attack crop predators. Research is ongoing in growing agricultural products directly in the laboratory using genetically engineered bacteria. Also envisioned is a major commercial role for genetically engineered plants as chemical factories. For example, organic plastics are already being produced in this manner (Borlaug, 1997).

Genetically engineered animals are being developed as living factories for the production of pharmaceuticals and as sources of organs for transplantation into humans. (New animals created through the process of cross-species gene transfer are called xenographs. The transplanting of organs across species is called xenotransplantation.) A combination of genetic engineering and cloning is leading to the development of animals for meat with less fat. Fish are being genetically engineered to grow larger and more rapidly. Could such a bounty be questioned? What happens when a genetically modified (GM) plant starts to grow wild in nature, its seeds spread by the wind? Will harm come to the birds that eat its seeds?

The topic of agricultural biotechnology used in the production of food for human consumption is an important issue for technology education to address. One facet of this issue is the fact that food is viewed as an entitlement in most of the developed world. In the United States, Europe, and other parts of the world, food is abundant. This abundant food supply is increasingly taken for granted. On any given day, the United States has less than 45 days of food supplies, and these supplies are viewed as a "surplus." In contrast, a 200-day supply of oil is maintained and it is viewed as a "strategic reserve." Thoughtful reflection about these facts might lead one to question the underlying ethical priorities. Each person participating in a technological world has the responsibility to be educated about the social responsibility of such attitudes. The task of developing technological literacy includes raising public awareness about the economic significance of an adequate, safe, and affordable food supply.

Failure to raise public awareness can be traced to two main problems. First is the economic benefit of improving nutrition and the difficulty to measure food safety. How does one place a value on the improved nutritional content of food or on food that is safer to eat? Second, there is the problem of how food safety issues are reported. A few people dying from food poisoning is newsworthy, whereas preventing millions from ever running such a risk is not (Hoban and Kendall, 1992).

## TACO BELL AND THE CONSUMER'S RIGHT TO KNOW

Food labeling should provide accurate information, based on scientific facts, not commercial prejudice. Labeling foods differently, simply because they are genetically engineered, is argued by some to be prejudicial, but others contend consumers have the right to know all the facts about the source of food products. Jerry Caulder, Chairman of the Xyris Corporation, a San Diego-based biotechnology firm, argued in an informal discussion that overwhelming scientific evidence suggests that genetically engineered food is no different than nonengineered food. However, Mockhiber and Weissman (2000) presented evidence that severe disruptions can occur when products not labeled as genetically modified or approved for human consumption reach the market and are then publicly identified.

The discovery that millions of pounds of corn in the United States were contaminated by being mixed with genetically altered corn not approved for human consumption can create havoc for the biotech industry. Genetically Engineered Food Alert, a coalition of health, consumer, and environmental groups, announced on September 18, 2000, that testing of Taco Bell brand taco shells revealed that the tacos contained Cry9C corn, marketed by the French biotechnology company Aventis under the name StarLink. Taco Bell brand taco shells sold in grocery stores are made by Kraft Foods, Inc., a subsidiary of Philip Morris. Since the September 18 announcement, Genetically Engineered Food Alert has identified other taco brands that contain StarLink corn, including tacos sold by Safeway and Western Family. Food companies have now spent hundreds of millions of dollars to remove contaminated corn from the food supply. StarLink corn is spliced with a protein that kills insect pests. The U.S. Environmental Protection Agency approved StarLink in 1998 for use in animal feed or nonfood industrial purposes only. It withheld approval for introduction into the food supply on the grounds that it did not have satisfactory data to show it would not trigger allergic reactions.

1. If you were managing the process of removing unapproved food products from the market, what would you do with them?
2. A charitable organization requests that the food products being removed from store shelves be given to them for distribution to homeless and hungry people. You and members of your family have previously consumed the products with no ill effects. The food products will be labeled to inform recipients about what they contain; the people being offered the food will choose whether or not to eat the food. How would you respond to this request?
3. What are your concerns about genetically engineered food products?
4. How would you balance concerns for product safety with economic considerations?

The complexity of the issues involving agricultural biotechnology can be further illustrated by the case of the amazing soybean. There are some 8,000 food products that are derived from soybeans that end up in the hands of consumers. Should we attempt to label each one individually because genetically modified soybeans may have been used as a raw base material? Furthermore, who should be responsible for such a task? Should we also label dairy products and milk derivatives, produced by cows fed with attrizine-treated corn? How do we judge what is safe and what is not? What rules do we use to guide us if not those of science? Is it socially undesirable to impose unreasonable costs on the consumer to provide food warning labels with little practical content (Macer, 1994)?

Ultimately, the success of agricultural biotechnology will be decided in the marketplace. Past experiences indicate that consumers can effectively sort through misinformation and decide about the real value of new goods and services. When color televisions were introduced, there were warnings about potential risks from "mutations." With microwave ovens, there were warnings about risks of "abnormalities" from radiation. The widespread adoption of both technologies suggests that when scare tactics lack solid scientific evidence, they tend to be short-lived. This will, no doubt, prove true for agricultural biotechnology as society increasingly becomes technologically literate.

## THE CASE OF ANIMAL CLONING

In early 1997, a research team in Scotland cloned a sheep named Dolly by modifying technology developed some decades previously with amphibians. Then, in July of 1998, researchers at the University of Hawaii produced mouse clones and developed a process by which mass cloning could occur. The technique used in both cases, somatic cell nuclear transfer, involves taking a nucleus from a somatic cell, placing it in an enucleated ovum, and implanting the ovum into a host uterus.

The cloning of Dolly brought to the forefront a longstanding debate about cloning human beings. The National Bioethics Advisory Commission recommended a 5-year moratorium on any attempts to create a child through somatic cell nuclear transfer in the United States and urged the President to work with all other nations to do the same. With the moratorium in place in the United States, legislative attempts to exercise permanent control over human cloning, such as the federal "Prohibition of Cloning of Human Beings Act of 1998," have been introduced in Congress.

Human cloning is a matter of concern for both the medical and biotechnology professions' attention because it would involve medical procedures and biological technology, and it may result in the creation of new genetic and psychological conditions that would require professional care. Therefore, not only society, but also the medical profession must evaluate the ethics of human cloning and, in particular, the potential role of physicians in any applications of cloning technology involving human beings.

1. Comment on the viability of a company that offered a service cloning people's pets.
2. If cloning and other biotechnologies could be used to remove genetic diseases and defects from farm animals, should these procedures be readily available and universally applied?
3. What concerns would you have about consuming food products produced from animals that had been cloned?

Agricultural biotechnologies must continue to be developed so that our ability to produce affordable, safe, high-quality food is enhanced. At the same time, the pressure on natural resources should be minimized. The government must also do its part. It must ensure the efficacy, quality, and safety of new products, while resisting the temptation to engage in social engineering. Consumers in the marketplace, rather than government mandates and regulation, can best answer questions of whether new products are desirable or needed.

Agricultural biotechnology is still in its infancy. Its real impact on the world's capacity to produce safe food to feed an expanding population, or its impact on natural resources, will not be fully felt for many years. Technology education should play a role in helping people to recognize the economic and social impacts of biotechnology as it seeks to prepare a technologically literate populace that is able to participate and make choices based on facts, not scare tactics or social prejudice. The balance between positive and negative impacts of agricultural applications of biotechnology remain a challenge requiring educated decision making.

## ETHICAL ISSUES IN PHARMACEUTICAL BIOTECHNOLOGY

Within the last decade, developments in biochemistry, molecular and structural biology, and microbiology have fundamentally altered pharmaceutical and biomedical research. These new applications of advances in bioscience have led directly to societal benefits through improved medical diagnosis and therapeutics. Biotechnology has become by far one of the most rapidly developing areas in pharmaceutical research (Evans and Relling, 1999).

Increasing proportions of multinational pharmaceutical firms' research and development budgets are devoted to biotechnology, and several of the biotechnology-based firms that were started only a few years ago are now evolving into fully integrated pharmaceutical companies. The intersection of growth within this biotechnology sector and social impact of the products produced has fueled concern over ethical considerations related to pharmaceutical biotechnology. A brief overview of the development of biopharmaceuticals is provided to describe the context for the case study in this section.

## SCOPE AND SCALE OF PHARMACEUTICAL BIOTECHNOLOGY

Eighty-four biopharmaceuticals are currently in general medical use and several are added to this list each day. Some 60 million patients worldwide thus far have benefited from these drugs, and the industry's global market value currently stands in excess of $12 billion. According to present estimates, around 500 biopharmaceuticals are currently undergoing clinical trials, ensuring that growth within this sector will continue (Walsh, 1998). Whereas early recombinant products approved were invariably replacement proteins, displaying amino acid sequences identical to the native human molecule (for example, human insulin, Factor VIII, and growth hormone), the proportion of engineered products being introduced is increasing. Major target indications of biopharmaceuticals currently undergoing clinical trials include cancer, cardiovascular disease, and infectious disease—the major killers within the developed world. Although in excess of 30 nucleic acid-based drugs (that is, gene therapy/vaccines, antisense, and ribozymes) are currently being evaluated, the majority are in the early stages of clinical development. The exponential growth in both the number of patients benefiting from biopharmaceuticals and the economic impact of the biopharmaceutical industry is well-documented, but questions of access to new treatments and the appropriate allocation of these technological resources remain at the center of public policy (prescription drug legislation) and are an ethical concern.

## THE EMERGENCE OF BIOPHARMACEUTICALS

Since the turn of the last century, biological research has been identifying biomolecules with therapeutic potential, several of which (for example, antibodies, blood products, and insulin) could be collected in quantities sufficient to facilitate their widespread medical use by direct extraction from the native source material. However, because most proteins are produced naturally in exceedingly low quantities, the extraction of medically useful amounts of many other biomolecules proved impractical (Akerele, Heywood, and Synge, 1991).

Although advances in chemical synthesis and semisynthesis provided alternative means of large-scale production of some biomolecules (for example, therapeutic peptides, some alkaloids, Paclitaxel [taxol] and Docetaxel [taxotere]), these approaches involved significant technical difficulties and economic costs, particularly for large proteins. Two discoveries, both reported in the mid-1970s, finally overcame these difficulties: (*a*) hybridoma technology (which facilitates the large-scale production of a monospecific antibody raised against virtually any antigen of choice), and (*b*) genetic engineering (which facilitates the large-scale production of virtually any protein after its amino acid sequence has been determined).

These discoveries prompted the establishment of dozens of biotechnology companies dedicated to producing therapeutic proteins. By the beginning of the 1980s, the biopharmaceutical era had truly begun. Recombinant DNA technology had a threefold impact on the manufacture of therapeutic proteins. Not only did it allow biotechnologists to overcome problems relating to source availability and product safety, but it also facilitated the development of modified protein medicines through protein engineering.

Problems of source availability due to low-level production by the native source have been eliminated. Even when donors could provide adequate supplies of biochemicals, collection was often difficult, unattractive, or even downright dangerous. A sample of the application of recombinant technology to produce biopharmaceuticals in modern treatment drugs and their prerecombinant method of production are compared in Table 3-3.

Table 3-3. Advantages of Specific Biopharmaceuticals Produced by Recombinant Means

| Biopharmaceutical | Prerecombinant Method of Production | Advantages of Recombinant Method |
|---|---|---|
| Interferon (IFNs) | Leukocyte IFN (mixture of $\alpha$-IFNs) produced by animal cell culture using the Namalwa (human lymphoblastoid) cell line. Other IFNs not available in medically useful quantities. | Makes specific $\alpha$-IFNs as well as IFN-$\beta$ and IFN-$\gamma$ preparations available in medically useful quantities. |
| Follicle-stimulating hormone (FSH) | Produced by direct extraction from the urine of postmenopausal women. Such "menotropin" preparations are still produced. | Facilitates production of FSH free from contaminating LH (luteinizing hormone), provides a source independent of urine. |

| | | |
|---|---|---|
| Hirudin | Small quantities purified from its native source (buccal secretion of the leech *Hirudo medicinalis*), but not available in quantities to allow medical application. | Makes hirudin available in clinically useful quantities, and production is independent of collecting leech saliva. |
| Ancrod | Small quantities of this anticoagulant can be purified from its native source, the venom of the Malaysian pit viper. | Makes ancrod available in useful quantities and without the need to "milk" snake venom. |
| Blood factors | Produced by direct extraction from human blood donations. Production by such means still ongoing. | Eliminates the risk of accidental transmission of blood-borne pathogens. |
| Human growth hormone (hGH) | Direct extraction from the pituitaries of deceased humans. | Eliminates the risk of transmission of pathogens such as the Creutzfeld–Jakob disease (CJD) causative agent. |

In other instances, recombinant forms of certain therapeutic proteins have gained a significant market share as a result of product safety concerns. An example of a significant concern that was raised involved the treatment of dwarfism by administration of human growth hormone (hGH) extracted from human pituitaries. This biotechnology came to an abrupt end in 1985 when a link between this treatment and transmission of Creutzfeld–Jakob disease (CJD) was discovered. Fortunately, a recombinant hGH product was just being made available (Protropin; Genentech, San Francisco, CA), and subsequently, several other recombinant hGH products have also gained approval. The potential for accidental transmission of disease has also served as the main rationale for the development and approval of several recombinant blood factors and also of hepatitis B vaccines.

## RAPID RISES IN BIOPHARMACEUTICAL PATENTS

Even though there are significant differences between the laws governing biopharmaceutical patents in the United States and in Europe, the number of patent applications for biopharmaceuticals has risen dramatically in the past decade on both sides of the Atlantic. In 1978, the United States Patent and Trademark Office (USPTO) received just 30 patent applications from the whole biotechnology field. By 1995, this figure had reached 15,600. In 2001, the technology center in the USPTO, of which the biotechnology examining division is a major part (about two-thirds of its examiners specialize in biotechnology), received 34,527 filings. The number of patents granted was also on the increase (USPTO, 2003). In 2001, according to statistics in Derwent World Patents Index (n.d.), 5,170 biotechnology patents were granted. An overview of the growth in patents granted by the USPTO for biopharmaceutical products is shown in Table 3-4.

Table 3-4. USPTO Biopharmaceutical Patents Issued by Year

| Year | Number of Patents |
|---|---|
| 1992 | 1,398 |
| 1993 | 1,532 |
| 1994 | 1,406 |
| 1995 | 1,588 |
| 1996 | 2,233 |
| 1997 | 3,148 |
| 1998 | 4,655 |
| 1999 | 5,054 |
| 2000 | 4,585 |
| 2001 | 5,107 |

Source: Derwent World Patents Index. Available at http://library.dialog.com/bluesheets/html/b10351.html.

## ETHICAL CHALLENGES RELATED TO BIOPHARMACEUTICAL TECHNOLOGIES

Now that scientists have a map of the human genome, a remaining question is how it corresponds to different diseases. The mapping of the human genome to corresponding diseases will allow medical scientists to understand each disease, in all its variations, at a genomic level. Scientists aided by technology will be able to determine how a disease is caused by a specific anomaly on a piece of DNA. Drug therapies will be targeted to work with an individual's particular genetic abnormality. But this kind of personalized medicine could have a significant impact on the pharmaceutical industry. With personalized medicine, a potential impact could be that treatments will be less severe and will have fewer side effects because they will be developed based on each patient's individual genetic map. The flip side, of course, is that we will need many, many more personalized treatments. This is contrary to current pharmaceutical targets. All of our medicines have been directed at 500 (disease) targets. Estimates indicate 5,000 to 10,000 targets in the future. On the face of it, this may sound like a great moneymaking opportunity for pharmaceutical companies, but this might not be the case (Swindells and Overington, 2002).

Today, drug developers make back the money they spend on research, development, and testing by targeting the most prevalent diseases. They make drugs for big groups of people, while those with a rare condition often wait in vain for suitable help. Drugs for rare diseases are called "orphans" for a reason; they are no one's primary concern. Pharmaceutical companies must make money to stay in business. It does not make sense financially to produce a drug that would only be prescribed for a small fraction of the population. This raises ethical issues, however, as those who suffer rare diseases have less chance of research and development finding a cure.

1. If you were the CEO of a major pharmaceutical company, what criteria would you use to determine whether to conduct research on a promising new drug?
2. Describe the responsibility, if any, for a pharmaceutical company to develop new products that would combat rare diseases.

The genetic revolution will involve ethical issues even more complex than those surrounding persons seeking treatment for rare diseases. New drugs will be developed for people seeking physical performance enhancements. The world of sports has already produced numerous examples of persons who sought to alter their physical attributes through use of steroids and other drugs. The ethical, moral, societal, and economic challenges of science and technology will continue to play a prominent role in the field of pharmaceutical biotechnology. How we negotiate, appropriate, apply, and control the products of pharmaceutical biotechnology will define who we are as a civil society well into the future.

## ETHICAL ISSUES IN ENVIRONMENTAL BIOTECHNOLOGY

Environmental biotechnology is defined as the development, use, and regulation of biological systems for remediation of contaminated environments (land, air, water) and for environment-friendly processes (green manufacturing technologies and sustainable development). An example of an application of environmental biotechnology can be illustrated in the case of polluted water and soil by heavy metals deposited in mining and industrial operations. This is an environmental problem in many areas around the world.

Although most plant species cannot tolerate high concentrations of heavy metals, certain plants thrive under these conditions. Of those species that actively accumulate metals, some called "hyperaccumulators" store metals in their tissues at concentrations far exceeding those in the environment. Hyperaccumulators of Ni, Co, Cu, Zn, Mn, Pb, Cd, Cr, and Se have been identified, and several hyperaccumulating species of Pb, Co, and Ni are native to Australia. Concentrations of metals found in these plants generally range from 1 to 5 percent measured on a dry weight basis; however, levels above 10 percent have been found in particular organs of some species. Hyperaccumulator plants have enormous potential for phytoremediation of contaminated land, being capable of translocating metal ions from the soil to the leaves, which can then be periodically harvested for disposal or metal recovery (Brooks et al., 1998).

In another example of environmental biotechnology, industrial and mining waste can be mitigated using bioorganisms. Australia is a bountiful country with regard to its mineral deposits, and, consequently, mining is an important industry there. Traditional mining technologies are, in many cases, being gradually replaced with more economically and environmentally friendly processes using microbial-mineral interactions or "bioleaching." Microorganisms not only have amazing potential to take up heavy metals from mine waste, but also can be used to remove contaminants produced by other industrial processes, such as lead-acid battery manufacturing and chrome plating (Baker et al., 1994).

## THE CASE OF SEWAGE SLUDGE DISPOSAL

Sewage sludge disposal is a serious worldwide problem. Because of increased environmental awareness and stringent environmental standards governing the disposal of sewage sludge (set by different environmental protection agencies), its utilization in agricultural production has been gaining increasing interest and attention in recent years. When properly processed, land application of sewage sludge offers economic and nutrient recycling advantages over the traditional disposal options, such as incineration for dry sewage and sea disposal (Hernandez et al., 2002; U. S. Environmental Protection Agency [USEPA], 2000). Nevertheless, potential risks related to the accumulation of heavy metals and organic compounds, as well as pathogen contamination, must be taken into consideration.

1. Comment on the use of sewage sludge as a fertilizer for land being used to grow vegetables.
2. What is the desirable balance between self-regulation and government oversight for industries processing sewage sludge for use as fertilizer?

With respect to direct environmental applications of biotechnology, there are developments as well. An example of replacing a chemical process with a bio-based one comes from Kenya and a product called BIOFIX, used as an alternative to chemical fertilizers. Measurements show that 100 g of BIOFIX replace 90 kg of chemical nitrogen (Zechendorf, 1999). This reduces not only toxic environmental exposure, but also potential for direct harm to humans because fertilizers are spread manually in most developing countries.

In the energy sector, Brazil has explored a strategy for using soy oil and sugarcane as a source of ethanol to be refined into biodiesel (Ford, 2000). Realistically though, any effort to promote bio-based substitutes for petroleum products will have to be undertaken specifically to support either a particular primary production sector or in response to environmental directives because oil continues to be available at relatively low costs.

When thinking about the uses of biotechnology and the environment, a significant ethical question can be asked. Should the environment be protected because of intrinsic moral values or should the environment be protected because it is a valuable resource for mankind? The answer to this question will impact support for environmental biotechnologies that might be beneficial to the biosphere but produce limited economic returns.

## ETHICAL ISSUES IN INDUSTRIAL BIOTECHNOLOGY

Industrial biotechnologies can be grouped into three main categories: (*a*) industrial supplies (biochemicals, enzymes, and reagents for industrial and food processing); (*b*) environmental (pollution diagnostics, products for pollution prevention, and bioremediation); and (*c*) energy (fuels from renewable resources). Industrial processing refers to chemicals, pulp and paper, textiles and leather, whereas environmental applications relate mostly to products for the metal and mineral industries. In the energy sector, an important area is the replacement of fossil fuels with renewable raw materials (Griffiths, 2001; Kate and Laird, 1999).

*Ethics and the Design and Development of Technological Systems*

The expected economic benefits from the industrial biotechnology sector are orders of magnitude smaller than those found in pharmaceuticals and agriculture (Griffiths, 2001; Kate and Laird, 1999). This essentially limits activity to current niche markets and relatively minor production changes in industry settings. The capital outlays required to use bio-based industrial processes are presently higher than those of traditional mechanical or chemical processes, so it can be difficult to justify starting new industries based on biotechnology. This means such innovations are likely to be incorporated into existing industries at a very slow rate.

## IMPEDIMENTS TO INDUSTRIAL BIOTECHNOLOGIES

From a practical perspective, only countries with existing industries and some scientific capacity can realistically consider applying biotechnology to industrial processes. Early studies in industrialized countries have shown that replacing existing chemical-based production methods can lower operating and energy costs (Griffiths, 2001). This research also revealed a consistent reduction in hazardous waste products. Thus, the industrial application of biotechnology, if managed properly, could have a net positive effect. The qualifier "net" is an important one in this context. The potential negative consequences of worker displacement, dependence on new materials, and changes in process management should also be considered.

The implementation of industrial biotechnology also faces financial barriers. Largely, this comes down to not having enough money for capital equipment upgrades. The related constraints include lack of private venture capital and underdeveloped stock market resources to invest in yet unproven technologies.

Structural resistance can also hinder the successful introduction of biotechnology into current industrial enterprises. One source of resistance can come from chemical and mechanical engineers with no biological training and minimal knowledge of the potential benefits of biology to their trades. They might not see any obvious advantages to changing processes. To effectively manage this internal resistance, there would need to be a local pool of scientists to explain the role of biotechnology in the manufacturing process and participate during the technology transfer process.

Another potential source of structural resistance comes from the nature and context of the people in the targeted industry. Senge et al.

(1999) argued that the challenge of fear and anxiety in people is the most frequently faced challenge in sustaining profound change and the most difficult to overcome. Furthermore, to be used effectively, processes involving biotechnology require industries and people to apply specific production standards. It is important to bring industrial processes involving biotechnology up to international standards from the beginning, but finding or training the human capital to do so is a significant barrier.

## ETHICAL CHALLENGES IN INDUSTRIAL BIOTECHNOLOGY

Taking the conditions previously discussed into consideration, it seems that the most likely form of industrial biotechnology that could be introduced in the near future is in cleaner production processes. This type of technology introduction would target specific parts of a given production process, could show economic benefits relatively quickly, and would also help companies conform to environmental obligations and social trends favoring "green" products. One factor that should be remembered is that often "dirty" industrial processes, particularly in foreign countries, arise from the tendency of foreign investors to engage in ventures with high yields in the short-term (Kryl, 2001). This raises the ethical challenges that are at the center of sustainable development, manufacturing, and green building. It is often foreign-owned firms that need to be convinced of the value of cleaner production. Adoption might best be encouraged by less emphasis on biotechnology being "clean" per se, and more promotion of the environmental and economic sustainability that flows from producing wastes that are degradable or recyclable (Griffiths, 2001).

In the short-term, industrial and environmental biotechnology will likely only benefit those countries that have already established some industrial enterprises that need to meet a set of environmental regulations. Reasons for this include the reality that it is generally easier to modify existing industries than to create new ones and that even minimal pollution control measures can be used to convince industrialists to take advantage of external assistance, either national or foreign, to adopt a bio-based cleaner production process.

1. If you were the CEO of a company with international investors, to what extent would you implement industrial biotechnologies that might reduce short-term profits while benefiting the environment?

2. Describe an appropriate balance between responsibility to investors and concern for the environment that will be inherited by future generations with regard to the environmental impact of industrial processes.

## SUMMARY

In this section, bioethical issues related to risk, choice, the environment, agriculture, pharmaceuticals, industrial biotechnology, genetic engineering, and equity were discussed. Consideration was also given to who should make decisions, rights of consumers, and developing and maintaining trust. Decisions related to biotechnologies should be made with an eye on previous "revolutions" in agriculture and industry and the ways values were challenged as a consequence. In retrospect, the genetic engineering revolution might not appear to have been particularly special, but until the fears of the public are appropriately addressed, the topic will continue to be controversial. Only by continuing fundamental research on risks and benefits of transgenics will there be adequate information to determine the usefulness of genetic engineering in agriculture, gene therapy, treatment of diseases, environmental restoration, and industrial and economic development.

There are many lessons that can be learned from public attitudes toward applications of biotechnology. Although risk assessment of new technologies has traditionally been based mainly on a risk versus benefit analysis, recent attention has been given to public perceptions in the development of biotechnology-related goods and services. A major tracking study examining public attitudes toward biotechnology and gene technology was conducted in July 2001 for the Australian Department of Industry Sciences and Resources program named Biotechnology Australia. The tracking study found increased analysis of both industrial processes and outcomes. The public was increasingly interested to know who would benefit from a new technology; would it be the community or a company? They also wanted to know if the processes of development would be harmful to humans, animals, and the environment. Products and processes that failed to indicate community benefits over company profit and did not demonstrate environmental benefits over harm were judged to be unethical (Commonwealth of Australia, 2001). In general, applications that were not seen to benefit humanity in a practical and ethical manner resulted in negative impressions. Most respondents perceived applications offering only cosmetic benefits as particularly negative.

One of the most challenging areas of concern related to ethics and biotechnology involves use and modification of genetic material in human cells. Focus group data about the exploration of modifying genetic material in human cells has shown that acceptability of this biotechnology hinges on how the application could be used or its outcome. If the application was able to improve human health, people tended to view it as an ethical and acceptable application of biotechnology. Applications leading to other outcomes were viewed less favorably.

As concerns about genetically modified crops, biotechnology, and technology transfer have come to the forefront of media coverage and precipitated response in the form of governmental policies, it has become clear that all citizens in this technological world could be impacted. To combat uncertainties about biotechnology and technology transfer, firsthand knowledge of these technologies must become part of education of every child. If citizens are to participate in the debate over issues, questions, and topics related to biotechnologies, technological literacy will be an essential prerequisite.

# ENERGY AND POWER TECHNOLOGIES

*Len S. Litowitz*

Technologies associated with energy and power encompass a wide array of resources that support life in a technological world. From electricity to gasoline, people are dependent on external sources of energy for almost everything they do. Numerous areas of energy and power production and use involve ethical issues, but perhaps no part of that field has precipitated more controversy than that of nuclear power. In this section, consideration is given to a case study involving a nuclear power technology and some of the related ethical issues of which technologically literate citizens should be aware.

Breeder reacting is a technology that is probably not familiar to most Americans. Even so, this technology could extend the amount of fissionable uranium that is suitable for use as fuel in nuclear power plants by a magnitude of about 50. The technology works by taking nonfissionable uranium and transforming it into plutonium, an element that can be used as fuel in most present-day nuclear power plants. Because more than 99 percent of all uranium ore is nonfissionable, the ability to convert this uranium into a fissionable fuel source could extend the nuclear generating capacity of this nation and allow for substantial expansion for many decades to come. With breeder reacting, more usable energy is produced by the reactor while in operation than is consumed.

Imagine burning a ton of coal along with another noncombustible substance such as iron ore in a generating plant, only to be able to remove that iron ore from the boiler and use it as fuel similar to coal at another generating plant. That is the essence of a breeder reactor. It is a proven technology that is seldom discussed, yet it is a technology that has been available for more than 50 years. It is even used successfully by other countries, including allies such as France and Japan. Where did this seemingly wonderful technology originate? Right here. It was American nuclear scientists that invented the first breeder reactor in the 1940s. So why don't we use breeder reacting in this country? That is a question that leads to one of the great ethical dilemmas associated with energy and power technologies of the present era.

## A BRIEF HISTORY OF BREEDER REACTING

The initial use of the breeder reactor had nothing to do with power generation. In fact, it was a technology that was invented for war, creating one stigma that breeder reacting has yet to outlive among some people. The process of nuclear fission was discovered in 1939. Up until that time, the process of splitting atoms had eluded scientists in both the United States and in Europe. By 1942, a small reactor constructed at the University of Chicago had sustained a brief chain reaction proving that if one atom could be split, the byproducts would split other atoms while simultaneously giving off a tremendous amount of energy in the process.

As World War II raged in Europe, it was widely known that German scientists were also working on achieving nuclear fission. The peacetime applications of nuclear power were placed on the back burner, and an all-out effort was mounted to create a nuclear bomb. To do so would require complex extraction methods to separate fissionable Uranium 235 (U235) from its much more popular cousin, Uranium 238 (U238), or the construction of a breeder reacting facility capable of producing highly enriched plutonium 239 (Pu239). Hedging their bets, the scientists on the project chose to pursue both methods of fuel production for constructing nuclear bombs, and both methods worked. The bomb that was dropped on Hiroshima used almost pure U235, whereas the bomb that was dropped on Nagasaki used breeder generated Pu239. Despite the horrific destruction caused by the bombs, many envisioned nuclear power and breeder reacting as a means of providing clean and inexpensive electricity for decades to come. The breeder served as a model for experimental reactors that paved the way for the first commercial reactor to generate electricity in Shippingport, Pennsylvania, in 1957.

## ETHICAL CONSIDERATIONS

The technology chosen for the development of commercial reactors used U235 as a fuel, not Pu239. In addition, the reactor design selected was such that it was not capable of producing large amounts of plutonium. The breeder reactor technology that had initiated the birth of nuclear power for both war and for peace was mothballed in favor of a nuclear-generating technology that was viewed as inherently safer. The decision not to pursue breeder reacting in the United States was as much an ethical decision as it was a technological decision. It was based in part on the fact that Pu239 is a human-made element with a half-life of over 24,000 years. This is considerably longer than the half-life of U235, its radioactive cousin. The extensive half-life means that spent fuel must be isolated from society for literally thousands of years. In addition, Pu239 is the fuel most commonly associated with nuclear bombs because it takes less plutonium than uranium to produce a bomb. Potential use of plutonium in power-generating reactors raised concerns about the use of nuclear power fuel or generation byproducts to produce nuclear weapons.

Other countries, including France, Japan, and Russia, chose to pursue the development of breeder reacting to produce plutonium, whereas the United States did not. Their reasons for pursuing breeder-reactor technology are equally as compelling as the reasons that the United States did not pursue the technology. After all, the successful development of a breeder-reactor program virtually ensures a limitless supply of electricity to any country that can build a nuclear reactor and has access to U238 to breed into plutonium. The process creates a usable fuel out of a previously unusable element. Furthermore, Pu239, when used as fuel for power generation, does not produce any fossil fuel pollutants, thus helping to reduce acid rain and gases that contribute to the greenhouse effect.

It is not unusual for legislators to seek expert testimony as they determine which laws to pass and how to budget tax revenues. In many instances, both sides of an issue are presented and the advantages and disadvantages are weighed as decisions are made. An important part of technological literacy is the ability to present rational arguments that can persuade others to manage resources and implement technologies wisely. Within this context, communication skills are an important complement to technical expertise.

1. Prepare and deliver a five-minute presentation either in favor of or opposing the implementation of breeder-reactor technology in the United States.
2. In preparing a rebuttal for arguments for or against the use of breeder reactors, what key points should be considered?
3. What resources might be used to learn more about breeder reactors? How can the accuracy and reliability of these sources be verified?

One of the ways technological literacy regarding energy and power can be enhanced is through the use of classroom debates in technology education classes or other related courses. Students working in small groups of two or three can review literature to develop the case for or against a selected topic. There are other energy and power production methods besides breeder reactors that are equally as controversial, so students in an entire class could experience researching and developing a case for or against a particular method. Table 3-5 illustrates how arguments on both sides of an issue can be developed. After being provided with adequate time for preparation, student groups can present their arguments in a structured classroom activity. The following guidelines can be used to manage this process:

1. The group that will present first (pro or con) is determined by coin flip. They are referred to as Group A.
2. Both groups assigned to a particular energy source are allowed five minutes of preparation time at the front of the classroom side by side.
3. When the presenters for Group A indicate they are about to start, the timer starts a clock. Group A is allowed five minutes to make a logical argument for (or against) the energy source that they have been assigned. Then, the students in Group B provide their five-minute presentation.
4. When both teams have presented, Group A is allowed two minutes for rebuttal; then Group B is allowed two minutes for rebuttal.
5. An additional three minutes are allowed for questions from the audience (the remainder of the class) to conclude the debate session.

Table 3-5. Samples of Arguments For and Against Selected Energy and Power Technologies

| Pro | Con |
|---|---|
| Solar energy represents a clean and inexpensive alternative to fossil fuels that could provide vast amounts of energy to this country if research in solar technology is appropriately funded. | Solar energy is so widely disbursed and the apparatus required to collect large amounts of energy is so expensive that it is not worth pursuing as an energy source. |
| Biomass fuels could reduce our dependence on foreign oil and help to move this country from an oil-based economy toward a more divergently energy-dependent economy. | Biomass fuels simply do not yield the quality of energy per volume that is necessary to sustain a modern economy. |
| Syn-fuels like shale oil and tar sands found in the western United States represent a viable alternative to foreign oil. Their development will create thousands of new jobs serving to\ bolster the domestic economy while easing our dependence on imported oil. | Syn-fuels are too difficult to retrieve and process so as to be economical. Because they are difficult to retrieve, conventional energy sources such as oil would have to triple or quadruple in price before syn-fuels could be competitive with imported oil. |

Classroom debates can provide for lively and educational experiences and could be used following presentation of any of the case studies in this chapter. Each topic should take about one half-hour in total so an entire class could participate within a couple of class periods. A sample scoring rubric is provided in Table 3-6. It is intended to be modified for individual classroom use. For instance, if an Exploring Technology course had completed a unit on communications, the scoring rubric might be modified to include a category on visual aids. One category that should not be modified significantly is the ethics category. The reason is simple. If the study of ethics is expected to be emphasized, it should be present in the scoring rubric, and a thorough explanation of ethical considerations should be rewarded with a higher score.

## SUMMARY

Ethical issues related to energy and power are complex and multidimensional. Environmentally friendly energy and power technologies are rarely the least expensive means of producing large-scale power. Other technologies such as breeder reacting that might make good economic

*Ethics and the Design and Development of Technological Systems*

Table 3-6. Sample Scoring Rubric for Classroom Debate

**Presentation Developed a Logical Argument**
    Logical and factually accurate ................................................................. 26–30
    Factually accurate but not logical ............................................................. 21–25
    Logical but not factually accurate ............................................................ 16–20
    Not logical or factually accurate .............................................................. 0–15

**Presentation Addressed Ethical Considerations**
    Multiple ethical considerations addressed ............................................. 16–20
    At least two ethical considerations addressed ...................................... 11–15
    At least one ethical consideration addressed ........................................ 5–10
    Ethical considerations not addressed ..................................................... 0–4

**Presentation Addressed Multiple Aspects (in addition to ethical considerations)**
    Addressed economic, technological, social, and other considerations ........ 16–20
    Addressed at least three different aspects ............................................. 11–15
    Addressed at least two different aspects ............................................... 5–10
    Only addressed one significant aspect .................................................... 1–4
    Did not address any significant aspects .................................................. 0

**Presentation Included All Group Members**
    All group members participated equally .................................................. 10
    Two out of three group members did most of the presenting ............ 8
    One of the group members did most of the presenting ...................... 6
    Group did not present on the assigned date ......................................... 0–4

**Presentation Cited Multiple Resources**
    Included at least five credible references ............................................... 10
    Included at least four credible references .............................................. 8
    Included at least three credible references ............................................ 6
    Included at least two credible references ............................................... 4
    Included at least one credible reference ................................................ 2
    Did not include any credible references ................................................. 0

**Presentation Fit the Timeframe**
    Filled the allocated time windows exactly ............................................... 10
    Filled the allocated time windows within 30 seconds .......................... 8
    Filled the allocated time windows within 1 minute .............................. 6
    Did not fill the allocated time windows ................................................. 0–4

Total Points _____ (100 points possible)

sense carry the baggage of long-term environmental concerns and even the possibility of nuclear proliferation. The nuclear industry as a whole has fallen on hard times as a result of fear from accidents like the Three Mile Island incident in 1979, the terrible explosion of the Chernobyl nuclear plant in Russia in 1986, inexpensive fossil fuel prices throughout the 1990s, and a failure to satisfactorily resolve nuclear waste issues. Even so, many people associated with the energy industry view the future use of additional nuclear power, and even breeder reacting, as inevitable. It will be important for technologically literate citizens to have input into these decisions and ethical as well as technical factors should be considered. The fact that a technology is possible and economical is not enough. Long-term impacts on the environment and quality of life must be considered to protect the well-being of future generations.

## INFORMATION AND COMMUNICATION TECHNOLOGIES

*Mark Sanders*

Much of the promise of the "information age" has come to fruition with remarkable haste. Not long ago, cartoon character Dick Tracy wore a science fiction videophone on his wrist. Dwight Eisenhower, President of the United States from 1953 to 1961, had advisors who imagined a "memex" machine that would mimic our brain's ability to instantly associate one idea to another idea. "Big Brother" was a fictional knowledge repository who knew everything about everybody, and the head of IBM said the world would only have need for a handful of computers.

We've come a long way in a short time. Outlandish ideas are now our reality—a reality that comes packaged with a host of new ethical issues and questions. The Obmnibus Communications Act of 1934 could not foresee the information age and the myriad of attending ethical issues. Relatively few have a solid grasp of the technical understandings of our new information technologies, making it difficult to develop related policy and legislation. Only in very recent years have our legislators attempted to address this new wave of information age ethical dilemmas.

The Omnibus Communication and the Digital Millennium Copyright Acts passed in the last year of the twentieth century are a first attempt at updating communication legislation crafted before the invention of the computer. We have a long way to go with respect to understanding and addressing the ethical issues relating to our new information and communication technologies. Technology education has an important role to play in this regard.

The following case studies illustrate some of the ethical issues citizens in a technological world must be prepared to resolve. Embedded in these circumstances are choices that have to do with preserving the environment, personal quality of life, and individual privacy. These are typical of the types of ethical concerns resulting from innovations in information technologies.

## PERSONAL COMMUNICATION SYSTEMS POLLUTION

In the 1990s, cellular telephones took America and the world by storm. People began to wear their phones like clothing. Conversations were no longer tethered to the home; all of a sudden, they were everywhere. At the turn of the century, more than 100 million people in the United States alone owned mobile phones. According to the Cellular Telecommunications and Internet Association, that number would double by 2005 (*Cell Phone Waste*, 2002).

Unfortunately, portable conversations are not without their drawbacks. Some people have begun to question whether wireless technologies are causing health risks. In 1996, the World Health Organization undertook a 10-year research project to study the impact of electromagnetic fields (EMFs) on humans and the environment. They were concerned enough to recommend the establishment of international standards for radio-frequency exposure. Other EMF studies have thus far been inconclusive regarding these health risks; medical experts have recommended erring on the side of caution.

Factors such as the environmental costs of portable communication systems have also been identified as concerns. Cellular phone towers—which must be located high on the horizon in order to function—are a source of visual pollution. More troubling is the substantial and rapidly increasing quantity of toxic waste resulting from the uncontrolled disposal of "old" cellular phones—each of which contains batteries and other materials that are unfriendly to our environment.

A 2002 study conducted by Inform, an environmental research organization, found that most cell phones are kept for about 18 months and are then often discarded with household trash. With as many as 500 million "old" cellular phones stashed in drawers around the home and office waiting to be trashed, the potential damage to the environment is very significant. The estimated 130 million discarded cell phones represent about 65,000 tons of toxic waste. When we add other portables, such as pagers, handheld computers, and music players to the disposal picture, the potential long-term environmental damages of these portable communication systems are alarming (*Cell Phone Waste*, 2002)!

1. What is the current extent of this problem?
2. Who should take responsibility for this problem? The cell phone manufacturers? The consumers who purchase cell phones? Federal, state, or local governments?
3. Do individuals have an inalienable right to portable communications, even if they result in damaging the global environment that everyone shares?
4. How is this problem exacerbated by the disposal of laptop and desktop computers? Research and summarize the data on these environmental costs.

## YOU'VE GOT SPAM

In 1969, the U.S. government funded DARPANET, a computer-based communication network for the U.S. military, and by 1972 the system included 37 computers. The network got a big boost in 1984, when the National Science Foundation funded NSF Net, which connected five university computing centers to assist with research efforts and related communications. By the late 1980s, access had spread to most universities, but it was the 1993 development of Mosaic—the first widely available Web browser—that pushed the Internet to all corners of the world.

Then, the inevitable happened; entrepreneurs, large and small, realized the Internet could be used to promote anything and everything to everyone, everywhere. From all corners of the world, they began mass e-mailing advertisements to every e-mail address they could put their hands upon. After all, it was free.

Normal e-mail "traffic" on the Internet increased exponentially, but "spam," as the mass marketing e-mails became known (named for the distasteful product served to American soldiers in World War II), traffic increased at an even greater rate. Spam was predicted to account for the majority of e-mail as early as July 2003.

"Spammers," the purveyors of spam, typically purchase their e-mail lists for as little as $25 per million addresses from e-mail brokers who "harvest" these addresses from public places on the Internet. Thus, everyone who uses the Internet is victimized by spam. Frequent Internet users—for example, those whose work requires Internet use—can be overwhelmed by this problem. A recent study by the Federal Trade Commission (FTC, 2002) found that 86 percent of the addresses posted to Web pages and/or newsgroups received spam.

The low cost of e-mail has encouraged all kinds of unwanted e-mail in addition to commercial spam, such as jokes innocently circulated among friends and illegal get-rich-quick chain letter scams endlessly circulated by unsuspecting individuals. Pornography spammers—who push images as well as text on the unsuspecting, have caused many to raise first and fourth amendment issues regarding our rights of free speech and privacy.

In the United States, the FTC has tried to address spam problems. In addition to discouraging chain letters—which are illegal—the FTC offers spam-reduction strategies and fields complaints about spammers from the public. The prediction that spam will become the majority of all e-mail in 2003, however, suggests the FTC strategies are not working.

Everyone who uses electronic mail in their work is spending a substantial and increasing amount of time sorting unwanted spam from their e-mail docket. The considerable amount of time required of nearly every worker nearly every working day to cull out these unwanted messages represents untold millions of wasted hours and dollars annually.

Despite the seriousness of the problem, spam is proving to be a difficult problem to solve. The Internet is a global network of international networks and is, therefore, difficult to govern or regulate. For example, as difficult as it might be to pass and enforce anti-spam legislation in the United States, doing so would likely impact only those spammers operating in the States. Curtailing the problem may require the acceptance and enforcement of spamming regulations and legislation in virtually all countries throughout the world, a truly profound task.

With no end currently in sight, the spam problem poses a significant threat to the effectiveness of the Internet as a communications medium. The Internet, which enjoyed explosive growth throughout the 1990s as it became the most efficient communication system in history, is now ironically threatened by its own success.

1. If on New Year's Day you send a joke to two friends, who, in turn, each send that joke to two additional different friends, and so forth like that for the rest of the month—how many people will have received that joke by February 1st?
2. Given the remarkable answer to the preceding question, what ethical obligations do individuals have to limit the amount of electronic mail they personally distribute?
3. What fourth amendment rights (to privacy) do individual e-mail users have with respect to spam?
4. Comment on rights of spammers in the United States to distribute unsolicited e-mails to millions of individuals via the e-mail system.
5. Describe the extent to which first amendment rights (to free speech) should protect spammers if the spam being sent promotes pornographic materials and includes pornographic images in the body of the e-mail note.
6. Given the global nature of the Internet, what strategies might be used to address the serious and rapidly growing problem of unwanted spam?

## THE DELICATE BALANCE: FREEDOM, PRIVACY, AND SECURITY

When terrorists brought the World Trade Center towers down on September 11, 2001, it changed—perhaps forever—the delicate balance between privacy and security in America. The 9/11 tragedy left Americans wondering how such an attack could possibly take us by surprise on our home soil. Americans wondered how it could be possible that many of the terrorists had been living in the United States and plotting the attack for years, and yet our government was completely unable to put the pieces of the puzzle together in time to avert the disastrous events of that tragic day.

The bulk of the blame fell on our system of collecting and analyzing information critical to national security. This system of gathering "intelligence" was not working very effectively. Our government immediately went to work to fix the system.

Government entities such as the Central Intelligence Agency (CIA) had been dedicated to ferreting out intelligence in foreign lands. That system of information gathering routinely involved government-paid spies and covert operations. Because these strategies had always been used abroad, Americans and government officials rarely questioned the covert nature of the tactics employed.

Locating and monitoring prospective terrorists living in America called for new levels of surveillance and intelligence gathering right here in the "land of the free." President George W. Bush declared a global war on terrorism, and efforts to fix the "intelligence" problem began in earnest. Orwell's "Big Brother" vision—a central knowledge base that knew everything about everyone—was looking increasingly desirable to many Americans in the aftermath of September 11th.

Americans had grown accustomed to taking their freedoms and privacy for granted. Phone conversations, e-mail correspondence, banking transactions, and personal belongings were all considered very private matters before the 9/11 tragedy. Following terrorist attacks in New York City and Washington, D.C., it seemed most Americans became willing to forego a measure of privacy for the benefit of our national security.

Airport security was an immediate case in point. When the airports reopened after the tragedy—even in small airports—National Guardsmen with M-16 rifles stood watch at flight terminal gates. Passengers and their carry-on luggage were thoroughly searched. All questionable items found, including such items as knitting needles, nail files, and scissors were confiscated by the thousands each day.

Racial profiling became an issue. Many questioned whether persons of Arabic heritage or of the Muslim faith—whether American citizens or not—should be allowed to work in airports, travel freely about the country, or attend universities. Some commercial pilots refused to fly if a passenger with Arabic features boarded the plane. In short, race discrimination raised its ugly head in many ways. At some level, the freedoms of American citizens, foreign students studying in American universities, and others were compromised—sometimes greatly compromised.

During the Gulf War a decade earlier (August 1990 to July 1991), poor communications among governmental agencies had been blamed for hindering national security. In that era, key intelligence resided on a centralized computing system—a practice for which the FBI was later widely criticized. By the time of the 9/11 tragedy, this centralized intelligence had been distributed to a "network of networks," and a new information infrastructure built pathways from one agency computer to another, so data could be shared almost instantaneously. This new infrastructure was believed to result in a more efficient system of communication during the war on terrorism in Afghanistan in the months following the 9/11 tragedy (Abrahamson, 2002).

But September 11 made it clear that this improved system of intelligence gathering and analysis still was not good enough, especially right here on American soil. The demand for greater national security led to the passage of the Homeland Security Act (HSA) in November 2002, establishing a new federal Department of Homeland Security (DHS). The DHS pulled together 26 different governmental agencies and established 17 new agencies (Department of Homeland Security, n.d.).

One of the key responsibilities of the new DHS is "Intelligence and Threat Analysis," which requires the DHS to "fuse and analyze intelligence and other information pertaining to threats to the homeland from multiple sources—including the Central Intelligence Agency, National Security Administration, Federal Bureau of Investigation, Immigration and Naturalization Service, Drug Enforcement Agency, Department of Energy, Customs Department, Department of Transportation, and data gleaned from other organizations" (Department of Homeland Security, n.d.). The purpose, of course, is to provide increased access to information deemed critical to national security for authorized governmental agencies.

The immediate demand for better intelligence caused government agencies to seek greater access to private information about American citizens. The HSA quickly accommodated this need. For example, the HSA protects Internet Service Providers who turn over subscriber information to the government in "good faith," even when a warrant hasn't been obtained (Homeland Security Bill, 2002). It also allows government agencies to run surveillance on e-mail without court approval during the course of a cyberattack. Public libraries are increasingly being asked to provide information, formerly considered private, about books checked out by specific patrons. Federal ID programs have been discussed and proposed in legislation so that all persons living in America would be increasingly identified using biometric measures, such as the use of fingerprints or retinal scans for individual identification (Sullivan, 2002).

Those concerned with civil liberties and fourth amendment rights to privacy are troubled by the loss of privacy resulting from increased access. They argue that just as airline travelers forfeited a degree of privacy in order to fly after the 9/11 tragedy, increased access to personal information results in "private" information becoming public.

In the first years of the twenty-first century, public battles began to be waged between those who wanted to enhance the use of surveillance and security technologies and those who sought to maintain rights and freedoms they felt were provided by the United States Constitution. New facial recognition technology was opposed by the American Civil Liberties Union based on research showing identification to be very unreliable. The Food and Drug Administration (FDA) decided not to regulate the implantation of computer ID chips in humans, as long as they were used for "security, financial, and personal identification or safety applications" (ID Chip's Controversial Approval, 2002).

Others sought to protect the right to privacy. Legislation was introduced in 2002 in the Senate that requires companies to get explicit permission from individuals before the company can collect or share this personal data. This legislation was intended to prohibit states from creating their own privacy regulations (ComputerWorld, 2002). Similarly, two Senators proposed a new "Privacy Commission" that would provide a forum for discussion about the use of new surveillance technologies (Dinan, 2002).

Many people have become concerned with high-tech approaches to surveillance, such as the FBI's "Carnivore" software, which is designed to collect all e-mail coming from a specific personal computer. The FBI defends this practice by pointing out that they must demonstrate "probable cause" before using Carnivore, thus limiting its use for the collection of "hard evidence," rather than "intelligence." With the continuing threat of terrorist attacks on American soil, however, there is growing concern that tools such as Carnivore will be used more and more liberally in the future, thus compromising rights and freedoms of American citizens.

The delicate balance between security, privacy, and freedom has been increasingly scrutinized. Agencies such as Privacy International, the

Electronic Privacy Information Center, and Reporters without Borders argue that actions taken by the United States and other governments around the world to combat antiterrorism threaten individual freedoms, civil liberties, and even the future of the Internet as we know it today (Report Shows, 2002; Thibodeau, 2002). The United States Senate recently passed legislation that would create a National Emergency Technology Guard, a group of volunteers who will try to prevent and, if needed, respond to cybersecurity attacks in the United States. Opponents cite potential for volunteer infiltrators to "do more harm than good" (Nando Times, 2002).

Meanwhile, electronic security is said to be a problem all across America. One report gave 14 of 24 of the largest government agencies a failing grade for electronic security and gave government an "F" grade overall (Lee, 2002). Studies have indicated that university campuses need to beef up security measures (Agency Raises, 2002). Another study found that 12 percent of American corporations reported major security breaches in 2001 (Lemos, 2002). Moreover, Web servers have been found to be more vulnerable than ever to security attacks (Chiger, 2002). The potential for cyberterrorism is very real. In October 2002, nine of the Internet's 13 "root servers" were flooded with 30 to 40 times the normal amount of traffic, causing seven of the nine servers to fail, and the other two to be temporarily ineffective (Roberts, 2002).

There is no right answer as to how the balance between national security and the rights and freedoms of American citizens should be struck. There are trade-offs either way. All that is clear is that the tragedy of September 11 put this national debate on our radar screens and here it will remain for years to come.

1. What is your opinion regarding the amount of privacy you are willing to relinquish in the name of increased national security?
2. How does one know if freedoms surrendered result in improved national security?
3. In your opinion, are surveillance/information-gathering technologies such as wiretapping or network "sniffing"—the process of collecting all information sent from a specific computer on a network—good, bad, or neither?
4. Identify a specific technological device, system, or strategy used for intelligence gathering that is currently being used. Describe how it

works. Identify and discuss some of the issues of security, privacy, and freedom that are inextricably linked to the use of this particular technology.

5. Identify an example in your local community that illustrates the interplay between security and freedoms or individual rights. What local examples are you aware of in which security has limited freedoms and rights or vice versa?

# TRANSPORTATION TECHNOLOGIES
*Myra N. Womble and Stephanie Williams*

Transportation is historically one of the most important types of infrastructure, a technology that developed worldwide. Archaeologists believe that the very first step toward human-made transportation began in either Mesopotamia or Asia, sometime around 4000–3500 B.C., with the invention of the wheel and eventually leading to development of mass transportation (Mitchell, 2001). People's desire for mobility gave birth in 1769 to the first vehicle of record to move under its own power. This vehicle, using a steam engine for power, was a military tractor designed by Nicholas Joseph Cugnot and constructed by M. Brezin (Bottorff, n.d.; Bellis, n.d.). Joining in this worldwide evolution, Oliver Evans received the first U.S. patent for a steam-powered ground vehicle in 1789, and Richard Trevithick's steam-powered road carriage was introduced in Great Britain in 1801 (Bellis).

The development of mass automobile ground transportation moved beyond the steam engine with the earliest electric engines built between 1832 and 1839. Robert Anderson of Scotland is credited with having built the first electric carriage. Electric cars used rechargeable batteries that powered a small electric motor. The batteries had to be recharged often and the vehicles were heavy and slow. However, by the early 1900s, the sale of electric ground vehicles was higher than the sale of all other types of cars in the United States.

Invention of the internal combustion engine paved the way for a new type of vehicle that continues to dominate the consumer market today. The efforts of many scientists, most notably Gottlieb Daimler in 1885 and Karl Benz in 1886, paved the way for the highly successful and practical gasoline-powered vehicles found on modern roadways. Steam and electric engines were abandoned for the internal combustion engine using gasoline, petrol, diesel, and even kerosene. Automobile ground transportation has certainly come a long way since the Cugnot vehicle built in 1769, weighing in at 8,000 pounds, with a top speed of 2 miles per hour and a tendency to tip over frontward.

The case studies that follow provide opportunities to consider some of the ethical issues related to transportation technologies. Both cases focus on automobile transportation technologies, specifically hybrid electric vehicles (HEVs) and sport utility vehicles (SUVs). Automobiles are very familiar to most people in today's technological world, but the related ethical issues presented in these cases are sometimes overlooked by people.

## ARE TWO BETTER THAN ONE?

Most people have heard of hybrid cars because they have been in the news a lot. In fact, most automobile manufacturers have announced plans to manufacture their own versions if they have not already introduced models within their product lines. However, some consumers still ask, "What is a hybrid car?" According to Nice (2003), "Any vehicle that combines two or more sources of power that can directly or indirectly provide propulsion power is a hybrid" (Section 2). Studying the development of these vehicles can help in understanding their significance.

The type of automobile consumers drive has a greater impact on the environment than any other consumer choice. In just one year, the average car emits nearly 730 pounds of hazardous chemicals, plus more than 10,000 pounds of carbon dioxide, a significant contributor to global warming. In addition, low gas mileage leads to increased drilling for oil, more spills from oil tankers, and more refineries polluting the environment and adding to the greenhouse effect (Hybrid Cars, 2002). In 1993, the HEV program began as a 5-year cost-shared partnership between the U.S. Department of Energy (DOE) and three American auto manufacturers: General Motors, Ford, and DaimlerChrysler. These auto manufacturers agreed to fabricate production-feasible HEV propulsion systems by 1998. They also agreed to produce first-generation prototypes by 2000 and market-ready HEVs by 2003.

The overall goal of the HEV program was to develop production-feasible HEVs with twice the fuel performance of similar conventional vehicles, with comparable performance, safety, and costs. The program's goals were later merged with those of the Partnership for a New

Generation of Vehicles (PNGV, 2002; Hybrid Electric Vehicle Program, n.d.). The PNGV was created in 1993 to identify improvements in fuel efficiency, emissions, and safety. In 2002, a new research partnership, the FreedomCAR program, was formed to develop cars and trucks that were cheaper to operate, pollution free, competitively priced, and free from imported oil (FreedomCAR, 2002).

The automobile industry has provided the United States with an essential product that few people can do without. Regrettably, this transportation infrastructure has become one that triggers emission of hazardous chemicals, oil spills, and petroleum refineries. Awareness of these technological impacts coupled with consumers' concerns about poor gas mileage and high prices have acted as motivators to business, industry, and government, resulting in the production of hybrid cars from manufacturers such as Honda, Toyota, and Chrysler. Even SUV manufacturers are getting in on the hybrid rush with Ford Motor Company planning to have what would be the first hybrid SUV, a gas-electric version of the Escape, on the market by early 2004. Likewise, General Motors Corporation plans to launch a hybrid SUV, the Saturn Vue, in 2005 (Shirouzu, 2003).

Consumers who are deciding whether to purchase a hybrid vehicle are often influenced by ethical issues related to this technology. In fact, this is a good example of a right-versus-right decision. The purchase price of a hybrid vehicle is several thousand dollars more than the initial cost for a comparable conventional automobile. Fuel consumption and pollution emissions are less with the hybrid, so benefits include lower operating costs as well as reduced environmental impact. On the other hand, if the car is being financed, higher monthly car payments might offset reduced operating costs. For some people, passing along a cleaner environment to future generations might be a priority, but concerns about how the batteries in hybrids will be disposed of when they require replacement might be an issue. When faced with these types of decisions, there is not a clear right or wrong answer, but technological literacy and the ability to resolve ethical dilemmas can help citizens cope with these kinds of decisions.

1. Comment on whether the benefits of HEVs (that is, cost, tax incentives, fuel efficiency, low emissions) outweigh higher initial purchase prices and maintenance costs.

2. What ethical concerns might lead a consumer to purchase a hybrid vehicle?

3. Two of the three manufacturers that first introduced hybrid vehicles were not a part of the HEV program with Department of Energy backing. Why would they have produced these vehicles?
4. Special lanes have already been designated on some highways for cars carrying two or more persons. Argue in favor of or against laws that would place further restrictions on passenger car traffic to encourage either car pooling or use of hybrid or alternative fuel vehicles.
5. Many believe America's energy security is threatened by its dependence on foreign oil. State a position on this issue and conduct Internet research to support your stance.

## IS BIGGER BETTER?

Sports utility vehicles saturated the United States' automotive industry following the end of Desert Storm when the oil prices dropped and supplies were plentiful. Classified as light trucks, SUVs were once used to carry heavy loads or navigate rough terrain, but have now become commonplace as passenger automobiles on our roads and highways. "For the purpose of fuel economy standards, Department of Transportation defines a light truck as any truck or truck derivative with a gross vehicle weight rating of 8,500 pounds or less" (Yacobucci, 2002). Sizes of SUVs have tended to become larger and larger, and there are several SUVs weighing more than 8,500 pounds. American citizens have done a 180-degree turn since the gas crisis of the 1970s and appear to be less concerned about fuel consumption. The majority of SUVs cover 16–21 miles per gallon. In addition, since 9/11, land transportation has become the preferred mode of transportation so travel by automobile and consumption of gasoline has increased.

The design criteria for many SUVs allows them to tow heavy loads and to go off-road. For many consumers, however, these expensive and visible purchases are driven by status, competition, and security. It feels good to sit high and look down at drivers in smaller cars, and the large size can lend a sense of security. The large size also results in high fuel consumption, which produces more pollutants and adds to the greenhouse effect problem. The Sierra Club, an environmental lobby group, has been especially critical of some of the largest SUVs and has even suggested that these vehicles, and requirements for large quantities of fuel, have contributed to the need for United States military actions in the Middle East (Wells, 1999). Sports utility vehicles and other light trucks are held to less stringent emission regulations under the Clean Air Act and to lower fuel economy standards under the Energy Policy and Conservation Act (Yacobucci, 2002), and some groups have lobbied for changes in these laws.

In response to environmental concerns, both General Motors and Ford have pledged to increase fuel efficiency by 15 percent and 25 percent, respectively, by 2005. Design engineers face making power sacrifices to achieve increased mileage. If this results in dissatisfaction on the part of consumers, automotive manufacturers will be pressured to find solutions that maintain their market shares.

Another concern related to the popularity of SUVs is the fatality issue in collisions between SUVs and cars. The United States Department of Transportation (USDOT) reported that 80 percent of the fatalities between light trucks and cars were car occupants. Lewis (1998) noted that in 1992, the decline in fatal traffic accidents stopped at the same time light trucks and SUV sales exploded. It is difficult to establish a causal connection, but the coincidence cannot go unnoticed. By 1997, the National Highway and Traffic Safety Administration (NHTSA) reported that 44 percent of new vehicles sold were light trucks and SUVs (Lewis). This agency reported rollover propensity and crash compatibility as two characteristics of SUVs and other light trucks that have affected fatalities.

Safety is a consideration that prompts consumers to invest in SUVs, but NHTSA statistics have indicated a propensity for these vehicles to rollover in accidents. Rollovers accounted for 37 percent of fatal crashes involving SUVs compared to a 15 percent rollover rate in fatal crashes involving passenger cars. In 1996, rollover crashes were involved in 53 percent of all SUVs occupants' deaths in single vehicle crashes (Friends of the Earth, n.d.). In collisions with cars, the height of an SUV often allows it to miss the passenger car's energy-absorbing bumper altogether, causing the passenger car to absorb most of the impact force (Lewis, 1998). Use of standardized bumper heights, to light trucks on equal footing with cars, has been proposed as a solution, but this poses design challenges. Manufacturers will likely resist uniform bumper heights unless required to implement them.

There are a variety of ethical issues related to the case of SUVs in the United States, but most consumers probably never give this much thought when choosing to make a new car purchase. It is another right-versus-right decision. If someone needs an SUV to periodically tow their boat to the lake, but the majority of driving involves one or two persons in the car, is the purchase of an SUV justified? Should people not be able to freely choose what type of vehicle they prefer? To what extent do individuals have a responsibility to make choices that will preserve the environment, especially when others around them are disregarding options that would conserve natural resources and reduce environmentally damaging pollutants?

1. How has increased fuel consumption in the United States affected politics?
2. How are designers and manufacturers addressing concerns about SUV safety and fuel efficiency? How can SUV manufacturers address safety concerns and still meet consumers' demands?
3. There is a call for fuel efficiency for SUVs; however, designers are concerned about losing power to meet this potential requirement. How should this concern be addressed?
4. Use a Venn diagram to illustrate similarities and differences between two SUV models. Go beyond the exterior by investigating their design.
5. Describe the Clean Air Act and the Energy Policy and Conservation Act. What roles do these Acts play in SUV design and production?
6. What amount of pollutants are passenger cars, hybrid cars, SUVs, and trucks producing? Compare and contrast your findings. Do your results support the fact that SUVs are under less-stringent emissions regulations? What research has been done to show the effects of SUVs on the environment?
7. Select four SUV models and determine the following for each model: (*a*) fuel tank capacity, (*b*) average miles per gallon, and (*c*) fill-up cost (use fuel octane ratings specified by the vehicle manufacturers).
8. How many SUVs are sold yearly? How much does an average-size SUV cost?

# MANUFACTURING TECHNOLOGIES

*Richard D. Seymour*

In today's global marketplace, every manufacturer attempts to establish a core competency that separates their organization from competitors. For instance, certain corporations are environmentally focused, whereas others are associated with superior quality. The hallmark of a few manufacturing firms is outstanding customer service or loyalty to a longtime clientele. Success in modern manufacturing is often linked to specific attention to product research, design, engineering, mass production, or marketing.

Naturally, all producers are under tremendous pressure to deliver a product of superior craftsmanship and value to consumers at a realistic cost. But numerous technical and managerial problems can hinder the design, production, and marketing processes. Ethical issues run a parallel path as each decision or action carries a "good" versus "bad" consequence. For example, a quick fix during the design phase might lead to a catastrophic failure during final assembly. In a similar way, rushing products out the door to meet a production deadline might compromise quality or lead to future liability issues.

Today, ethical issues are at the core of both the manufacturing sector and our system of public education. This often makes for an interesting combination when a course is based on modern production, such as when teaching a class called Manufacturing Enterprise or Production Systems. In these courses, participants are taught about responsible actions related to the design and production of a physical artifact. A significant goal of the course might be to instill accountability in the student manufacturing team as they function in a factory-style setting.

Numerous additional examples of ethical issues related to manufacturing technologies can be found in the headlines of news media on a daily basis. Whether in the response of manufacturers to concerns of police regarding the safety of fuel tanks in their cars or in the management decisions made by corporate heads regarding offshore factory operations, ethical issues permeate this sector of our technological world.

The following cases were developed based on experiences within the context of technology education courses related to manufacturing technology. By providing scenarios similar to those experienced by other students, there is an implicit goal of encouraging thoughtfulness about circumstances to which learners could relate. Research on contextual teaching and learning has suggested that this is an important facet of effective instruction.

## THE CASE OF THE DEFECTIVE CORDS

The following example occurred in a class in which the focus of the student-developed manufacturing system was an electrical extension cord, a relatively simple product for an advanced class. The design of the extension cord was attractive to members of the class as it featured a durable body, large handle, and a long, heavy-duty cord. During the course, approximately 25 units were to be produced using student designed jigs and fixtures, inspection gages, and so on. Being a spring semester class, many students planned to purchase one of the units, at just over material cost, as a Father's Day gift so they had a vested interest in the success of the project.

Several issues began to surface during the production planning and engineering work. A small group of students missed deadlines or failed to help teammates during the initial preparation, and this led to early frustrations within the class. With a strict division of responsibilities for developing workstations, certain teams became leery of their colleagues. Attempts to correct an attendance problem did little to increase the effectiveness of the class. Ultimately, the fabrication teams finished their efforts far better than the assembly team planners. Concern over the organization of the final assembly line grew with the approach of the scheduled production day.

As one might suspect, the component parts of the extension cord were produced to high standards. But sub- and final assembly work was not completed in an efficient manner, nor did the final products match any level of acceptable quality. Several units did work, with minor modifications, but the pile of defects far exceeded the number of "quality products." Several students expected a poor grade for their lack of attention to detail and miserable production efforts.

The instructor returned the failed products to all the students in the class, this time with everyone in attendance and with their full attention. The session started low-key, with a review of how certain segments of production went well, whereas others were a complete failure. The focus switched to who was going to buy the few good products and the individuals who would receive the units. The young manufacturers calmly learned about the families of their classmates, people who were loved and admired and treasured. A number of students mentioned their fathers as the most influential person in their lives. Others cited family devotion, sacrifices, and support that had enabled them to attend college.

The next question caught the class off guard, as a relevant issue was raised; namely, why did they (that is, the students) want to harm or possibly kill these loved ones? If distributed, each of the defective extension cords could have started a fire or possibly shocked the end user. In the worst-case scenario, electrocution of a family member was not out of the realm of possibilities. The students began to see a clear relationship between their limited efforts in a manufacturing class and the potentially disastrous consequences. A harsh reality was beginning to set in for students who had accepted little responsibility earlier in the course.

Naturally, planning and a team approach had taken a "back seat" during much of the class. Those are the ingredients of a poor final score in most classes, but a failing grade was becoming the least of the worries of the students. Ethical issues, mostly related to product safety, personal involvement, and liability concerns, suddenly confronted the production team. It took considerable effort to rework the numerous defects that had rolled off the manufacturing line, but most students participated in efforts to reassemble the products in a proper manner.

As a result of the experience, members of the class discovered how manufacturers must adhere to strict guidelines, especially when it comes to quality and individual responsibility. Poor initial performance and a lack of group coordination led to the entire class having to rework the extension cords. The real learning occurred, however, when the students faced the ethical dilemma of a failed production-based team. Students moved beyond concerns about grades and were challenged to think about manufacturing practices that could result in personal injury or death.

1. If the manufacturing defects in the class project had not been discovered, who would have been responsible for the personal injuries that might have resulted?
2. Would more learning occur in a class experience like that described or in a class in which everything worked smoothly and products were produced without defects? If the defects enhanced learning, should they be "designed" into the manufacturing experience?
3. Describe a strategy for solving product quality issues in a real manufacturing plant setting. To what extent would similar problems on an

industrial assembly line be handled from an economic perspective as compared to an ethical perspective?

## MAKING A FAIR PROFIT

A second example might further illustrate the challenges faced by manufacturing educators, this time when addressing the generation of profits. Manufacturing firms exist to make money, but how much is "too much" when profit margins are established? The topic comes up frequently in a Manufacturing Enterprise class as the budget for a specific product and an entire company is determined.

The central theme of a Manufacturing Enterprise experience is to develop, produce, and market a mass-produced item at a profit. In reality, the course involves role-playing as a team of students simulate the actions of a private or corporate enterprise. Students typically assist in the design and engineering work, plus help to develop a budget for a mock company. They also staff the production line and market the final products to friends and peers. The ultimate goal of the class is to maximize the profits from the production and marketing of a single line of products. At the same time, students learn about industrial organizations, teamwork, production work, and other topics relevant to manufacturing processes.

In an enterprise class, the budget is established early in the course. A break-even chart helps focus the financial picture. Then, students, in both leadership and company personnel roles, help to create the overall budget. Because the students are wearing multiple hats during the course, they quickly discover that they may financially benefit in a number of ways. The challenge is determining how to balance the expenses and income, thus creating sufficient profits without exceeding marketable pricing.

Ethics typically play a huge role in the budgeting process. Students who do not plan on buying stock in the company favor a high rate of pay and substantial sales commissions. Those who plan to purchase stock want a minimal payroll and few financial incentives, such as safety awards or production bonuses. All students favor an exorbitant selling price, often forgetting that they too might be a "customer" later during the term. Naturally, a balance must be established between the selling price and the financial commitments of the firm.

Unfortunately, because students perform so many tasks in a Manufacturing Enterprise class, it begins to resemble "insider trading" several times during the course. After all, if a production team leader realizes that quality will suffer later in the term, they might favor a higher payroll versus higher final dividends. That's another way of saying "take the money and run now." Likewise, the head of the financial team monitors the expenses of the mock company and is alerted to deviations in the monetary plans earlier than other student workers, investors, or company leaders. Sometimes, the financial team is accused of modifying the books to benefit individual students or the instructor.

Discussion over the class's budget is usually a give-and-take activity among the student leadership team and the instructor. A spreadsheet program can be used to assist the students in comparing different values, such as varying wage rates or levels of sales commissions. Students in classes have been known to say, "Let's not pay the students (that is, line workers) for production time because they have to come to class anyway." "If we can get the materials donated by someone, we can eliminate our production expenses." Note: Few corporations get their raw materials free! "We can charge twice that price because most of the sales will be to family members; they will buy whatever we make anyway." "Let's eliminate the commission and sales awards. Why should we have to pay people to sell the products anyway?"

1. How would you respond to some of the suggestions just listed?
2. Where do most of the ethical issues in a Manufacturing Enterprise course arise—in the design and production processes or in managing the financial aspects of the project?
3. Who should profit the most from participation in a school manufacturing activity—production workers, student managers, the teacher, customers, others? Who should profit most from manufacturing activities in the corporate world?

## SUMMARY

A Manufacturing Enterprise course introduces the basics of production-based firms, but it can also highlight corporate greed, internal corruption, a poor work ethic, and other less-than-desirable traits. The teacher must stay ahead of developing attitudes and mannerisms to use the evolving course scenarios in an instructive way. The best instructors encourage the students to do the right thing, yet also outline the consequences if a negative or illegal course is followed.

Manufacturing education can be as simple as producing a perfect extension cord or developing a realistic budget for an enterprise. It is important to remember, however, that raw materials and a computer spreadsheet program do not have a personality or conscience. Young students must be taught how to assemble components or information in an acceptable manner. Both of the scenarios presented here required young learners to combine their upbringing, morals, knowledge, and sound judgments in an ethical manner.

Perhaps a central issue is determining the best time for stressing ethics as related to modern manufacturing. Is it preferable to cover the topic early in the course, or wait until something goes right or wrong? The best answer is YES! Students should become aware of proper and responsible actions at the start of each manufacturing class or unit. It is also helpful to review the correct procedures near the end of the course, especially if a major situation has arisen. The extension cord scenario is a prime example of how students can learn an important lesson about teamwork and quality from what appeared to be a failed manufacturing activity. When technology educators are mindful that technological literacy is more than technical skills with tools and processes, experiences involving both failure and success can be used to equip students to be good citizens in a technological world.

# CONSTRUCTION TECHNOLOGIES

*Jack W. Wescott*

The content area of construction in technology education has been previously defined as "the efficient use of manufactured goods, materials, and resources to build a structure on a site" (Wescott, 1994, p. 184). In addition, the study of construction in technology education has traditionally focused on the materials and process related to the design, construction, and use of structures (Henak, 1994). It should be noted, however, that many of the activities associated with construction technology require judgment, respect, and trust among the community of people participating in and affected by the act of constructing a structure. When issues related to judgment, respect, and trust arise, they are ethical issues that are not clearly defined by traditional construction practices, contracts, and laws. This section provides some guidelines for educators who are addressing ethical issues within the context of construction technologies, and includes some case studies illustrating ethical dilemmas that can arise in that field of work.

## CONSTRUCTION AND ETHICS

According to Wassman, Sullivan, and Palermo (2000), "the process of designing and constructing structures with the presumed intention of improving the quality of life implicitly requires judgment of the right thing to do" (p. 17). It is this underlying intent that links the study of ethics to construction technology.

The literature related to the study of ethics in construction technology includes three major classifications of ethical issues. Although these classifications are not meant to be all-inclusive, most ethical issues in construction technology can be classified as relationships, construction processes, and societal or cultural issues.

With regard to relationships, construction can be described as a social, political, economic, and cultural event. It is not a solitary process because it involves groups and communities of people committed to conceiving, designing, and constructing. In creating human habitats and other structures, construction processes are supported by concepts and practical knowledge of technology, history, theory, cultural heritage, and visions of the future.

As previously indicated, construction practices are not solitary in nature. Simple observation of a construction site indicates that individuals work in groups and these groups interface with other groups. Specific examples of relationships in construction include interactions among clients, structure users, subcontractors, materials suppliers, interior designers, landscapers, and the general public. Although many of these relationships are formed through a formal set of contracts and legal agreements, others are developed through governmental regulations or public law, such as code enforcement. Still others are formed as a result of a handshake on the construction site.

Ethical issues related to construction processes include an expectation that constructors remain current in the core knowledge and skills related to construction technology. They have an ethical responsibility to possess and exercise competent expertise in the construction activities they undertake. Numerous ethical issues are naturally embedded in the "processes" of construction, such as estimating and bidding, selection and purchase of materials, fabrication of structural components, and the adherence to codes and regulations. In addition, these ethical issues are often dealt with under stressful conditions due to the competitive nature of construction and the financial risks involved.

Beyond the processes listed previously, structures need to be built safely to protect people and goods from the forces of nature. Therefore, legal and moral obligations exist with regard to contracts, structural considerations, barrier-free accessibility, and quality.

Other ethical obligations in construction originate in the processes of participation in community affairs with respect to such issues as the environment, community planning, socioeconomic concerns, and public services. It can be argued that to be completely ethical, it is important to include the community and societal interests in decisions related to construction projects, even when they are not identified as a concern by the client. This infers that beyond the constructor's character and moral virtues, there is a set of values and concerns that lead to a commitment to the immediate geographic region and the local community.

The entities related to community and society that are affected by a construction project typically include (*a*) the immediate client or owner, (*b*) the users of the structure who often have little or no input to the design and construction process, (*c*) the general public whose quality of life is affected, and (*d*) the environment whose ability to regenerate and sustain humankind is affected by the interruptions associated with the construction process.

In summary, ethical issues related to the three areas of construction can be organized into the following areas: (*a*) formal and informal interaction and relationships with clients, subcontractors, architects, engineers, and others; (*b*) the mastery of construction knowledge and skills and the competent exercising of professional knowledge and judgment; and (*c*) the conduct of the contractor in relation to the immediate community and society with respect to such issues as the environment, community planning, socioeconomic, and public services.

The following list is a sampling of the ethical issues that occur naturally in construction technology. Furthermore, daily endeavors and decisions related to the construction process are driven by the following ethical decisions:

1. Business choices (marketing, deciding on which projects to undertake, which clients to work with, and so on)

2. Design deliberations and critiques (function, aesthetics, concepts)
3. Budgets (durability of construction, value for cost)
4. Client and subcontractor interactions (honoring contracts, fairness, trust and advising clients)
5. Contracts (equitable conditions, providing value for service fees, mutual respect, and duties)
6. Public presentations (who has the right to know and be advised about projects; who has input to design)
7. Development and recognition personnel

Although the preceding list is presented under the guises of business and professional construction practices, embedded within them are ethical questions. The following is a sampling of such ethical questions:

1. What are the motives, values, and intentions of potential clients? Do we agree with their values?
2. Who are the people who will be using the structures we design? How are they served?
3. Who and what are impacted by the project and in what ways?
4. What type of project is it? Is the project's purpose one that we could support?
5. Do we honor contracts that we enter into? Are we fair toward contractors and consultants?
6. Do we give proper credit to those whose talent and work efforts contribute to the work that is shaped?
7. Do discussions of aesthetics during design give rise to consideration of ethics? If so, how?
8. Do we advise our clients or simply honor their requests? Are advising and guiding clients professional "duties," or do constructors merely "serve" clients?
9. Are constructors "professionals"? Do members of a profession have special ethical duties and responsibilities?

## THE PLEASANT CITY JAIL

The following theoretical case study describes a real-life construction problem that contains examples of ethical issues previously described.

The ABC Construction Company is submitting a bid for the construction of a maximum-security prison in Pleasant Valley, Indiana. ABC is aggressively pursuing the contract because of the financial profit to be made. The project is also important to the company because ABC is located in Pleasant Valley, and the majority of the employees are residents of the area.

A statewide planning committee recently conducted a study to determine a location for the prison. It was the recommendation of the committee that the most appropriate location for the prison would be in the city of Pleasant Valley. This appeared to be a logical choice for the planning committee for several reasons. However, the main reason was the financial benefit to the community. Several manufacturing companies in the area had recently closed and moved operations to foreign countries. The committee implied that the construction and operation of the prison would provide over 1,200 jobs for the large number of unemployed residents. Also, the construction of the prison means that the state would contribute additional tax dollars to the city to support schools and other public service projects.

Another reason for being selected was that the city has a vacant 20-acre tract of land available that was formerly the town's landfill. Pleasant Valley is located near the middle of the state and is accessible by state and local roadways. A medium-sized residential subdivision and school are less than a mile from the proposed site, and that would make a convenient commute for the prison employees. Currently, the vacant property is only used as a temporary refuge for the overpopulated Canadian geese when they migrate in the fall.

1. Problem: What are the ethical decisions that the ABC Construction Company must address before it commits to the project?
2. Project type: How do you feel in general about state-supported government complexes (or other types of low-income housing, military installations, youth opportunity centers)? Are these types of structures appropriate for the needs of the community? Also, does the company ask their employees to work on projects that are against their personal beliefs?
3. Who the client is: Would you perform services for the state or federal government if you held personal values that differed from the agency's political, social, or religious position? In this case, should

personal values associated with the state corrections system have a significant impact on decisions related to the project?

4. Social-economic issues: What are the benefits of a construction project to the residents and the community as a whole? More important, it is often necessary to weigh the values of one ethical issue against another. For example, in the preceding case study, the contractor is forced to decide if new jobs and the economic well-being of the community are more valuable than the issues related to the environment. A second ethical decision is the risk of a maximum-security prison located in close proximity to a housing development and school. To what extent should local residents' fears of a prisoner escaping and causing harm to members of the residential development be considered?

5. Environmental issues: Is it appropriate to build the new structure on a site that was a former landfill? In addition to the concerns related to the construction process, it may also have long-term effects on the inmates and employees of the prison. Do we know what types of refuse were disposed of in the landfill? Did the local manufacturing companies that have since left the area use the landfill to dispose of chemical wastes? Should we be concerned about the Canadian geese during their fall migration? After all, there appears to be an overpopulation of Canadian geese in the area.

## SUMMARY

Because the case study presented is fictitious, we will never know the decisions of the ABC Construction Company. We can only speculate that there would be difficult decisions, and some of these would involve ethical issues. What is most important is that when choosing a course of action, constructors should exercise their professional knowledge, skill, and judgment toward positive ends being certain to bring professional knowledge, judgment, and fairness to bear on the important ethical issues of a construction project.

It is also helpful to recognize in what ways the ethical issues related to construction can be applied to other technologies that are a part of the designed world. The issues raised by the preceding example lead to similar, more broadly based questions regarding the benefits of technologies versus personal values and ethics. When we further examine the dilemma of ABC Construction Company, a series of more encompassing questions could be asked. Is a type of project ethically good or questionable?

One might believe that socially redeeming construction projects such as housing for the homeless, temporary housing for disaster victims, schools, day care centers, or hospitals would not be questioned from an ethical perspective. Even in these types of structures, it is important to determine the ethical dimensions of the construction project. For example, what are the motives of the client? How many people are affected by the project, and how are they affected? Is the purpose of the project restrictive or controlling? What are the social impacts on the users and the community, in general? Does the project really represent the interests of the populations it claims to serve? Are ecological concerns addressed?

Many of these same questions could be asked about the other technologies discussed in this chapter. Some make the assumption that technology is neutral. When technology is put into use, however, it is never neutral. These applications always have aspects related to ethics and often involve trade-offs and choices that can be characterized as right versus right.

Throughout this chapter, the various technologies that are a part of the designed world have been discussed. Topics related to medical technologies and biotechnologies were treated at length because they are relatively new to the content addressed in technology education classes. In some sections, case studies were taken from real life, and in others, hypothetical scenarios were developed to illustrate ethical issues in technology. In all instances, the intent was to enhance technological literacy by demonstrating the significance of ethics in our technological world. If citizens are not equipped to make good ethical decisions, their technical expertise might be of little benefit to themselves or others. In many ways, the very future of the world will be determined by these decisions that reflect ethics or a lack thereof.

# REFERENCES

Abrahamson, L. (2002, November 27). National Public Radio (live broadcast).
Agency raises the bar on tech security. (2002, February 27). *USA Today* (cited in Edupage 2/27/02).
Akerele, O., Heywood, V., & Synge, H. (Eds.). (1991). *Conservation of medicinal plants.* Cambridge, UK: Cambridge University Press.
Aronson, S. M. (2000). A wondrous thing, a ray of hope. *Medicine and Health, 83*(2), 34-35.
Baker, A. J. M., McGrath, S. P., Sidoli, C. M. D., & Reeves, R. D. (1994). Possibility of in situ heavy metal decontamination of polluted soils using crops of metal-accumulating plants. *Resources, Conservation, and Recycling, 11*(1-4), 41-49.
Battelle, P. (1976). Let me sleep: The story of Karen Ann Quinlan. *The Ladies Home Journal, 93*(9), 69-76.
Beach, M. C., & Morrison, R. S. (2002). The effect of do-not-resuscitate orders on physician decision-making. *Journal of the American Geriatrics Society, 50*(12), 2057-2061.
Bellis, M. (n.d.). The history of the automobile. Retrieved May 18, 2003, from http://www.inventors.about.com/library/weekly/aacarssteama.htm
Benner, P. (2003). Avoiding ethical emergencies. *American Journal of Critical Care, 12*(1), 71-72.
Bennett, I. J. (1977). *Technology as a shaping force. Doing and feeling worse: Health care in the United States.* Daedalus, 106(1), 125-133.
Bhardwaj, M. (1999). Ethical issues of Human Genome Project. Conference on Global Ethos, United Nations University, Tokyo, Japan. Retrieved February 4, 2003, from http://vulab.ias.unu.edu/GlobalEthos/papers/minakshi.html
Blake, R. H. (1988). *Life, death, and public policy.* DeKalb: Northern Illinois University Press.
Borlaug, N. (1997). Feeding a world of 10 billion people: The miracle ahead. *Plant Tissue Culture and Biotechnology, 3,* 119-127.
Bottorff, W. W. (n.d.). *What was the first car? A quick history of the automobile for young people.* Retrieved February 18, 2003, from http://www.ausbcomp.com/~bbott/cars/carhist.htm

Bronzino, J. D. (Ed.). (1992). *Management of medical technology: A primer for clinical engineers.* Boston: Butterworth-Heinemann.

Bronzino, J. D., Smith, V. H., & Wade, M. L. (1990). *Medical technology and society: An interdisciplinary perspective.* Cambridge, MA: The MIT Press.

Brooks, R. R., Chambers, M. F., Nicks, L. J., & Robinson, B. H. (1998). Phytomining. *Trends in Plant Science, 3*(9), 359-362.

Callahan, D. (2002). Slippery slope—Medical technology and the human future. *The Christian Century, 119*(20), 30-35.

Casey, L. B. (1976). A statement on the case of Karen Ann Quinlan. *The Catholic Mind, 74*(1301), 8-12.

*Cell phone waste.* (2002). Retrieved December 7, 2002, from http://www.cbsnews.com/stories/2002/05/08/tech/main508346.shtml

Chiger, S. (2002, April 22). Privacy groups fight for chat room rights. *ComputerWorld.* Retrieved April 22, 2002, from http://www.idg.net/ic_851985_1794_9-10000.html

Commonwealth of Australia. (2001). *Australian biotechnology report.* Paragon Printers. Australia: Canberra. Retrieved September 4, 2002, from http://www.biotechnology.gov.au/library/content_library/BA_BiotechnologyReport2001a.pdf

ComputerWorld. (2002, April 22). Retrieved January 15, 2003, from http://www.computerworld.com/securitytopics/security/hacking/story/0,10801,75336,00.html

Cruse, J. M. (1999). History of medicine: The metamorphosis of scientific medicine in the ever-present past. *The American Journal of the Medical Sciences, 318*(3), 171-180.

Department of Homeland Security. (n.d.). Retrieved December 23, 2003, from http://www.whitehouse.gov/deptofhomeland/sect1.html

Derwent World Patent Index. (n.d.). Retrieved January 23, 2003, from http://library.dialog.com/bluesheets/html/bl0351.html

Dickerson, S. S. (2002). Redefining life while forestalling death: Living with an implantable cardioverter defibrillator after a sudden cardiac death experience. *Qualitative Health Research, 12*(3), 360-372.

Dinan, Stephen. (2002, August 2). *Washington Times.* Retrieved August 5, 2002, from http://www.washtimes.com/national/20020802-473343.htm

Enderle, J. D., Blanchard, S. M., & Bronzino, J. D. (2000). *Introduction to biomedical engineering.* New York: Academic Press.

Evans, W. E., & Relling, M. V. (1999). Pharmacogenomics: Translating functional genomics into rational therapeutics. *Science 286*, 487-491.

Federal Trade Commission. (2002). Retrieved December 1, 2002, from http://www.ftc.gov/bcp/conline/pubs/alerts/spamalrt.htm

Fiechter, A., Beppu, T., & Beyeler, W. (Eds.). (2000). *History of modern biotechnology*. Berlin; New York: Springer.

Ford, S. (Ed.). (2000). Brazil plans sugar cane-soy biofuel to cut pollution. In *Sustainable development international: Strategies and technologies for local-global agenda 21 implementation*. Retrieved May 8, 2003, from http://www.sustdev.org/industry.news/092000/26.01.shtml

*FreedomCAR: The partnership to develop America's hydrogen economy of the future*. (2002). Retrieved August 17, 2003, from http://www.ott.doe.gov/fcar_partnership.shtml

Friends of the Earth. (n.d.). *Are SUVs safe?* Retrieved May 19, 2003, from http://www.suv.org/safety.html

Griffiths, M. (2001). *The application of biotechnology to industrial sustainability*. OECD, Paris. Retrieved May 8, 2003, from http://www1.oecd.org/publications/e-book/9301061e.pdf

Helwege, A. (1996). Preventative versus curative medicine: A policy exercise for the classroom. *The Journal of Economic Education, 27*(1), 59-72.

Henak, R. (1994). Rationale and structure of content for construction in technology education. In J. Wescott & R. Henak (Eds.), *Construction in technology education* (pp. 16-19). Peoria, IL: Glencoe/McGraw-Hill.

Hernandez, T., Moral, R., Perez-Espinosa, A., Moreno-Caselles, J., Perez-Murcia, M. D., & Garcia, C. (2002). Nitrogen mineralisation potential in calcareous soils amended with sewage sludge. *Bioresource Technology, 83*(3), 213-219.

Hoban, T. J., & Kendall, P. A. (1992). *Consumer attitudes about the use of biotechnology in agriculture and food production*. Raleigh: North Carolina State University.

Hoffman, B. (2002). Technological medicine and the autonomy of man. *Medicine, Health Care, and Philosophy, 5*(2), 157-167.

Homeland Security Bill Includes Internet Provisions. (2002, November 11). *New York Times* (cited in Edupage, 11/20/02).

Howell, J. D. (1996). *Technology in the hospital.* Baltimore, MD: The Johns Hopkins University Press.

Humber, J. M. (1991). Statutory criteria for determining human death. *Mercer Law Review, 42*(3), 1069-1085.

*Hybrid cars for a greener world.* (2002). Retrieved May 18, 2003, from http://www.networkforgood.org/topics/animal_environ/hybridcars/

*Hybrid electric vehicle program.* (n.d.). Retrieved May 18, 2003, from http://www.ott.doe.gov/hev/background.html

ID Chip's Controversial Approval. (2002, October 23). *Wired News.* Retrieved October 25, 2002, from http://www.wired.com/news/politics/0,1283,55952,00.html

International Technology Education Association (ITEA). (2000). *Standards for technological literacy: Content for the study of technology.* Reston, VA: Author.

Juma, C. (1989). *The gene hunters: Biotechnology and the scramble for seeds.* Princeton, NJ: Princeton University Press.

Kate, K., & Laird, S. A. (1999). *The commercial use of biodiversity.* London: Earthscan Publications.

Kohl, M. (1976). On death, dying, and the Karen Ann Quinlan case. *The Humanist, 36*(1), 16.

Kotzar, G., Freas, M., Abel, P., Fleischman, A., Roy, S., Zorman, C., Moran, J. M., & Melzak, J. (2002). Evaluation of MEMS materials of construction for implantable medical devices. *Biomaterials, 23*(13), 2737-2750.

Kristiansen, B., & Ratledge, C. (2001). *Basic biotechnology* (2nd ed.). Cambridge, MA: Cambridge University Press.

Kryl, D. (2001). Environmental and industrial biotechnology in developing countries. *Journal of Environmental and Biotechnology, 3*(4). Retrieved November 5, 2002, from http://www.ejbiotechnology.info/content/archive.html

Lalan, S., Pomerantseva, I., & Vacanti, J. P. (2001). Tissue engineering and its potential impact on surgery. *World Journal of Surgery, 25*(11), 1458-1466.

Lee, C. (2002, November 20). Agencies fail cyber test. *Washington Post.* Retrieved November 21, 2002, from http://www.washingtonpost.com/ac2/wp-yn?pagename=article&node=&contentId=A12321-2002Nov19&notFound=true

Lemos, R. (2002, September 10). Survey: Security budgets on the rise. *CNET.* Retrieved September 12, 2002, from http://news.com.com/2100-1001-957364.html

Levinovitz, A. W., & Ringertz, N. (Eds.). (2001). *The Nobel Prize: The first 100 years.* London: Imperial College Press; Singapore; River Edge, NJ: World Scientific Publications.

Lewis, S. (1998, September 10). They just aren't compatible. *Machine Design, 70,* 76-79.

Macer, D. R. J. (1990). *Shaping genes: Ethics, law and science of using genetic technology in medicine and agriculture.* Christchurch: Eubios Ethics Institute.

Macer, D. R. J. (1994). *Bioethics for the people by the people.* Christchurch: Eubios Ethics Institute.

Macer, D. R. J. (Ed.). (2000). What the genome project means for society. In *Ethical challenges as we approach the end of the human genome project.* Christchurch: Eubios Ethics Institute.

Mauzur, D. J. (2002). Law and ethics: Trust as social capital and as encapsulated interests. *Journal of the Society for Medical Decision Making, 22*(4), 372-379.

Meyyappan, M. (2000). Nanotechnology—What is ahead? Aerospace Conference Proceedings of the Institute of Electrical and Electronics Engineers, *USA, 1,* 55-56.

Mitchell, E. (2001). *The history of transportation.* Retrieved May 31, 2003, from http://mnmn.essortment.com/transportationh_rgly.htm

Mockhiber, R., & Weissman, R. (2000, December). Enemies of the Future. *Multinational Monitor,* (21), 12. Retrieved May 8, 2003, from http://multinationalmonitor.org/mm2000/00december/enemies.html

Nando Times. (2002, November 21). Congress moves forward with technology guard. Retrieved November 21, 2002 from http://www.nandotimes.com/technology/story/473380p-3783273c.html

Nice, K. (2003). *How hybrid cars work.* Retrieved May 18, 2003, from http://auto.howstuffworks.com/hybrid-car1.htm

Ott, B. B. (1995). Defining and redefining death. *American Journal of Critical Care, 4*(6), 476-480.

*Partnership for a new generation of vehicles.* (2002, June 5). Retrieved May 18, 2003, from http://www.ta.doc.gov/PNGV-Archive/default.htm

Penticuff, J. H. (1990). Ethical issues in redefining death. *Journal of Neurosurgical Nursing, 22*(1), 48-49.

Polla, D. L., Erdman, A. G., Robbins, W. P., Markus, D. T., Diaz-Diaz, J., Rizq, R., Nam, Y., Brickner, H. T., Wang, A., & Krulevitch, P. (2000). Microdevices in medicine. *Annual Review of Biomedical Engineering, 2,* 551-576.

Rajput, V., & Bekes, C. E. (2002). Ethical issues in hospital medicine. *The Medical Clinics of North America, 86*(4), 869-886.

Reiser, S. J., Dyck, A. J., & Curran, W. J. (Eds.). (1997). *Medical ethics.* Cambridge, MA: The MIT Press.

Report shows increased government surveillance. (2002, September 3). *Associated Press* (cited in Edupage, 9/4/02).

Rifkin, J. (1998). *The biotech century.* London: Victor Gollanz.

Roberts, P. (2002). Major net backbone attack could be first of many. *InfoWorld.* Retrieved December 23, 2003, from http://archive.infoworld.com/articles/hn/xm1/02/10/23/021023hnoneofmany.xm1?s=IDGNS

Rosenblatt, J. (2000). *International conventions affecting children.* The Hague; Boston: Kluwer Law International.

Senge, P., Kleiner, A., Roberts, C., Ross, R., Roth, R., & Smith, B. (1999). *The dance of change: The challenges of sustaining momentum in learning organizations.* New York: Doubleday.

Shirouzu, N. (2003, February 6). When hybrid cars collide: Several gas-electric systems jockey to become a standard; Echoes of VHS vs. Betamax? *Wall Street Journal.*

Snellen, H. A., & Hollman, A. (1996). Willem Einthoven (1860–1927): Father of electrocardiography: Life and work, ancestors and contemporaries. *Medical History, 40*(4), 516.

Sullivan, B. (2002, May 2). Federal bill would require biometric driver's licenses within five years. *ComputerWorld.* Retrieved May 5, 2002, from http://www.idg.net/ic_855521_1794_9-10000.html

Swindells, M. B., & Overington, J. P. (2002). Prioritizing the proteome: Identifying pharmaceutically relevant targets. *Discovery Today, 7*(9), 516-521.

Thibodeau, P. (2002, September 4). Senate measure embraces opt-in: Class-action suits on privacy a possibility. *Wired News.*

U.S. Congress, Office of Technology Assessment. (1987). New Developments in Biotechnology, 2: Public Perceptions of Biotechnology-Background Paper (OTA-BP-BA-350). Washington DC: U. S. Government Printing Office.

U.S. Environmental Protection Agency. (2000). Alkaline stabilization of biosolids. Biosolids Technology Fact Sheet (USEPA Publication No. EPA 832-F-00-052). Washington, DC: Author. Retrieved May 8, 2003, from http://www.epa.gov/owm/mtb/alkaline_stabilization.pdf

U.S. Patent and Trademark Office. (USPTO). (2003). *Patent full-text and full-page image databases.* Retrieved May 8, 2003, from http://www.uspto.gov/patft/

Walker P. M. B. (Ed.). (1988). *Chambers science and technology dictionary.* Edinburgh: Chambers.

Walsh, G. (1998). *Biopharmaceuticals: Biochemistry and biotechnology.* New York: John Wiley and Sons.

*The Washington Post.* (1985, June 12). Karen Ann Quinlan dies at age 31; Coma case prompted historic ruling (p. A10).

Wassman, B., Sullivan, P., & Palermo, G. (2000). *Ethics and the practice of architecture.* New York: John Wiley and Sons.

Wells, P. (1999, April 5). Return of the good old super-guzzlers. *Alberta Report, 26,* 17-18.

Wescott, J. (1994). Preparing teachers for construction in technology education. In J. Wescott & R. Henak (Eds.). *Construction in technology education* (p. 184). Peoria, IL: Glencoe/McGraw-Hill.

Williams, D. (1999). Small is beautiful: Microparticle and nanoparticle technology in medical devices. *Medical Device Technology, 10*(3), 6, 8-9.

World Medical Association. (1999). Proposed revision of the Declaration of Helsinki. *Bulletin of Medical Ethics, 150,* 18-22.

Yacobucci, B. (2002). Sport utility vehicles, mini-vans, and light trucks: An overview of fuel economy and emissions standards (CRS Report for Congress RS20298). Washington, DC: National Council for Science and the Environment. Retrieved December 23, 2003, from http://www.ncseonline.org/NLE/CRSreports/air/air-32.ctm

Zajtchuk, R. (1999). New technologies in medicine: Biotechnology and nanotechnology. *Disease A Month, 45*(11), 449-495.

Zechendorf, B. (1999). *Trends in Biotechnology, 17,* 219-225.

# Ethics and the Assessment of Technological Impacts on Society

## Chapter 4

Robert C. Wicklein
The University of Georgia
Athens, GA

Technological progress impacting societies and cultures in the United States and around the world continues to advance at exponential rates. Scientists and technologists have designed systems to put people into outer space, clone animals, and provide telecommunications to all parts of the globe, but have failed to clean the slums of Calcutta or to elevate the standard of living of millions of people who live in abject poverty around the planet. As a whole, humans have become incredibly competent at developing new science and technology, but have failed to correct the age-old social and environmental problems that have plagued mankind for generations. As stated by Kristin Shrader-Frechette and Laura Westra (1997), "Humans' intellectual progress often outstrips their moral and ethical development" (p. 3). We are witnesses to a civilization that steadily loses ground in moral and ethical arenas while at the same time making incredible leaps in the advancement and development of science and technology.

Vanderburg (2000) speculated, "Modern civilization is lost in a labyrinth of technology created by its social and environmental implications" (p. xi). We have developed highly complex systems that churn out a myriad of high-tech devices while at the same time becoming more calloused and ignorant about the impact of these devices on society and the environment. Is there a solution to the burdens that science and technology place on society and the environment? Do we as a society have the intellectual and, more importantly, the moral fortitude to recast the way in which we go about developing our planet? Can we become more proactive in ways that prevent social and environmental blunders as we develop technology? The answer to these questions can be found only if we are able to understand the labyrinth of technology.

The motto of the 1933 Chicago World's Fair was, "Science Finds, Industry Applies, Man Conforms" (Norman 1993, xx). Although this motto would probably not be acceptable in today's politically correct culture, it may very well be accurate in describing how most industrialized societies function. It is not uncommon to hear people describe their lifestyles in relation to options that technology provides for them with little or no discussion about the limitations that technology places on them. For example, with the advent of wireless communication technology (a technology that has become pervasive in U.S. society within the past 10 years), many business activities and transactions have an integral tie to this form of technology. The average businessperson cannot imagine going to work without a cell phone or PDA. It would be a great impediment to conducting business if these technological systems failed or did not function correctly. The necessity of meeting with clients or coworkers face-to-face has been drastically diminished with the advent of these forms of technology. Regardless of how positively or negatively you may perceive this technology, it has had a huge impact on society and can be construed as an example of how "man" conforms to technology (remember business activities were performed successfully for centuries without wireless communication devices).

What impact have wireless communication devices had on society? Have wireless communication devices made business activities better? If so, how? Have wireless communication devices made business activities worse? If so, how? What impacts have wireless communication devices had on human relationships? How can we measure the impacts of wireless communication devices on society? These are the types of ethical questions that need to be asked as we consider the impacts that technology has on society.

The focus of this chapter is the role of ethics in assessing technological impacts on society and the importance of ethics in development of sustainable forms of technology. The narrative discusses the various forms of ethical thinking used to guide decision making in business, industrial, social, and political systems. In addition and as an application for technology teacher education, an examination (specifically Standards 4, 5, 6) of the *Standards for Technological Literacy* (International Technology Education Association, 2000) focuses on the principles and values that teacher educators should be exploring with preservice and in-service teachers with regard to technological impacts on society. The key points addressed in the chapter are (*a*) ethical worldviews—their impact on technology, (*b*) the role of technology teacher education in the teaching of ethics, (*c*) interpreting the *Standards for Technological Literacy* with regard to ethics instruction, and (*d*) a sample instructional activity.

The significance of this chapter is that it spotlights some of the significant issues and problems that are integral to understanding the ethics behind the technological impacts on society. Attention is directed at the root of ethical decision making as it applies to the development and utilization of technology to extend human capability.

## ETHICAL WORLDVIEWS— THEIR IMPACT ON TECHNOLOGY

There is almost a limitless number of worldviews (philosophies) that would address the issues related to ethics and the assessment of technological impacts on society. In Chapter 1, several of these approaches were presented along with a model for making difficult ethical decisions. Five worldviews are presented in this chapter to represent the underlying ethics for thinking about technological impacts on society. The five ethical worldviews are (a) skepticism, (b) cynicism, (c) relativism, (d) materialism, and (e) hierarchicalism. Each of these worldviews are discussed briefly.

### Skepticism

In this ethical worldview, there is no objective truth whatsoever; therefore, there are no rights and wrongs—only differences of opinion. Man is the measure of all things. Truth is totally dependent on the individual. The Sophist philosopher, Protagoras, first expressed this philosophy in the fifth and fourth century B.C. (Popkin, 1967); it combined a basic position of explaining the universe in terms of phenomenological aspects (for example, all I know is what I can see, hear, touch, taste, and smell) with personal experiences and interpretation. In this ethic, there are no universals—only subjective interpretations. Today, we see this ethical position presented in postmodern thinking and writing.

As applied to technological impacts on society, this ethic proposes that any form of explaining positive and negative technological impacts is acceptable based on a person's individual perspective. Tolerance of all viewpoints is a goal of this ethical position. Any judgments stating a collective or general evaluation of a positive or negative impact of technology on society are deemed inappropriate. For example, as an individual, I can declare that the application of automobile technology has a negative impact on me; however, I should not state that it has a negative impact on any other person or group.

## Cynicism

In this worldview, there are some objective truths and we can know them (for example, mathematics, physics), but there is no objective truth as it applies to subjective issues in ethics. The cynic philosophy was erratically developed and presented as early as the fourth century B.C. with its most notable advocate being Diogenes of Sinope (Kidd, 1967). The basic premise of the cynic worldview is that truth is only those things that can be proved through empirical research, and everything else is just subjective interpretation by an individual. In addition, the cynic deems that any issue that falls outside of empirical analysis is either pure foolishness or simply unworthy of further study.

The cynic worldview of the ethical impacts of technology on society are based only on scientific experimentation, such as the specific amounts of pollutants coming from automobiles combined with scientific experimentation showing the negative effects of these pollutants on plant and animal life. Individuals applying this ethical position could speak directly to this specific issue, but could not connect the same automobile with the social and emotional ramifications of a society that has splintered because of mobility brought on by the widespread use of automobile transportation. For the cynic, these types of ethical considerations are subjective and cannot be verified as truth because they are not supported by empirical data.

## Relativism

Relativism is an ethical worldview in which there is objective truth but there is no objective ethic (goodness). There is no real right and wrong, only individual will or desires (Kreeft, 1996). Moral or ethical relativism became widely accepted in the 1960s and is often presented as a choice between law and liberty (freedom—personal, sexual, emotional, psychological). The claims of moral relativism include:

1. All moral laws are man-made and are only custom or convention—not absolute.
2. There is no objective goodness—all values are relative, subjective, and personal.
3. There are no rights or wrongs—only individual wills, desires, and preferences.
4. Might makes right—for example, monarchy or democracy.

5. From the relativist perspective, to believe in moral absolutes is to join the unthinking majority who are traditionalists, "stuck-in-the-mud" conservatives, status quo, morally "hung-up," sexually impoverished yahoos.
6. Moral law is created rather than discovered, making morality more like art than science.
7. The only absolute is that all ethical truth is relative.

Philosophically, relativism can be explained in three different ways, each describing a structure in which this worldview may function within society (Brandt, 1967). In descriptive relativism, ethical disagreement follows along cultural lines. The culture is what defines the ethical system; therefore, the ethics within an Arab culture could be radically different from a European culture. According to the worldview of moral relativism, one ethic is no better or worse than another.

In metaethical relativism, ethical positions are neither true nor false but only express the attitudes of the speaker. Ethics are determined solely by an individual demonstrating or presenting a particular moral viewpoint. Because ethical truth is subjective, one point or position is of equal value with any other point or position.

Normative relativism is defined as ethical values that are based on the majority consensus (law of the land) within a given community, state, or nation. Ethics are based on the rules and regulations of a given government at a given time. Therefore, under this form of moral relativism (at least philosophically), the ethics of Hitler's Nazi Germany or Pol Pot's Communist Cambodia were no better or worse than those of Gandhi's India or Bush's United States of America.

The impacts of moral/ethical relativism on technology can be profound, but vary depending on which philosophical approach is applied. Based on the organizational structure of a given society (for example, elected representatives with the rule of law and appropriate checks and balances versus totalitarian regimes with little or no representation or checks and balances), the effects of an ethical worldview can be moderate to large in scope. End results of this way of thinking may give rise to the rapid expansion of technological systems or the rapid limitations of technological systems, depending on the culture and who is in control of the government. For example, when the Shah of Iran in the 1960s and 1970s decided to modernize his country through the revenues of massive oil exportation, it resulted initially in rapid technological growth both at the national and personal levels. Activities related to modernization and the subsequent affiliations with western cultures were viewed by certain segments within the Iranian society to be inappropriate and contrary to the Muslim religion. When the government of the Shah of Iran was overthrown in 1978, the relativistic worldview changed with regard to technological development, resulting in the rapid decline or limitation of technology. Technological innovations (or at least certain components of technology) were viewed as ethically "bad" or "wrong" based on the ruling authorities within the government.

When applied at the individual level, ethical relativism can take the form of people purchasing and utilizing technology to meet needs and satisfy wants based totally on a personal rationale. For example, an individual may decide to purchase a cellular telephone because he wants to be in constant touch with others. The impact of cell phone technology on personal relations is a debatable point but (and importantly) a moot point because within this worldview, there is no such thing as universal "rights or wrongs" or "good or bad." There are just personal and subjective choices. No advantage is realized by discussing the ethics of a technology such as cell phones because all have their own reasons and perspectives on the use and impact of this technology and each opinion is just as valid as any other.

## *Materialism*

Materialism is the term given to a family of doctrines concerning the nature of the world in which matter holds a primary position. Extreme materialism asserts that the real world consists of material things, varying in their states and relations, and nothing else. It is with extreme materialist views that we are here concerned. Thus, the cardinal tenet of materialism is "everything that is, is material—anything that isn't material, isn't real" (Campbell, 1967).

A material thing can be defined as being made up of parts possessing many physical properties and no other properties. The physical properties are position in space and time, size, shape, duration, mass, velocity, solidity, inertia, electric charge, spin, rigidity, temperature, hardness, and many others (more are being added as we discover new ways to measure things). These things make up the science of physics. In this philosophical worldview, the only thing that matters is what can be measured to verify its existence; if it cannot be measured, it either does not exist or is not worth exploration. Materialists have traditionally been determinists—there is a cause for every event (and the cause is based on physical sciences).

When applied to ethics, this worldview takes on a purely mechanistic form; ethical choices and decisions are the effects of matter relating to other matter, nothing more and nothing less. Personal choices are removed from this worldview and ethic; the only thing that matters is matter. Personal moral responsibility is also removed along with any form of guilt—something that makes this worldview highly attractive to many people.

As applied to the impacts on technology and specifically the teaching of technology, the application of the materialist worldview focuses exclusively on the mechanisms of a given technological topic. Instruction and learning often center on the internal operation of technology artifacts (for example, principles of the internal combustion engine, design of electrical circuitry, design and creation of Web sites). The primary goal is to communicate how technology works, how it is applied to perform various functions in business and industry, and what careers are connected with a particular technological system. Issues related to the values of the technology, impacts of the technology in nontechnical (nonmaterialistic) areas, and conceptual or philosophical dimensions receive little attention or are not considered at all. This approach is common in many technology education programs and, in the opinion of this author, is indicative of a significant weakness in the study and teaching of technology.

### *Hierarchicalism*

Up to this point, the ethical worldviews discussed have taken positions in which there is no specified universal normative ethic, meaning there are no absolutes that govern ethical actions (rights versus wrongs). A universal normative ethic is one that is prescriptive rather than descriptive. It is an ethic that commands certain courses of action as opposed to describing ethical actions. A normative ethic does not merely describe how people *do* act; rather, it prescribes how they *ought* to act. It is not an ethic of "is" but an ethic of the "ought" (Geisler, 1971, p. 21). The universality of the ethic expands its application by postulating that the prescriptive element of the ethic applies to all people in all situations regardless of time, place, and culture. In the ethical worldviews of skepticism, cynicism, relativism, and materialism, there are no ethical norms and there are no real ethical universals. In these worldviews, ethical decisions are based on and at the discretion of the individual, purely and simply. The issue of "ought" is not considered as a relevant factor when employing these ethical worldviews.

Hierarchicalism is unique from other worldviews in that it incorporates both ethical norms as well as ethical universals. In the hierarchical ethical worldview, there are many universal ethical norms, meaning there are ethical positions that are immutable and complete apart from any interpretation by people. For example, the hierarchical ethical position for truth telling or not committing murder is consistent in every circumstance at every time for every person with a notable exception. The hierarchical position holds that there are multiple ethical universals, but at various times these universals may conflict with each other, wherein one universal (for example, preventing murder) impacts on another universal (for example, telling the truth). When multiple universals come in conflict, the hierarchical ethic holds that the higher ethic is always accepted as the primary goal or focus of human action. Therefore, lifesaving is more important and more ethical than truth telling.

When applying this ethical position to technological impacts on society, individuals are charged with evaluating technological systems to consider the long-term as well as the short-term ramifications of these systems. For example, is it "right" to use technology to retrieve and view personal and private information about an individual? The issue of privacy is a huge concern in our modern societies; legislation has been introduced and passed into law at many local, state, and national levels (Annas, Glantz, and Roche, 1995) to limit unauthorized access to personal and private information. If, in a given situation, extremely private information (for example, genetic codes) about an individual is needed to prevent the death of an unconscious individual (same person), is it ethical to access these records without permission? According to the ethic of hierarchicalism, the saving of life is a higher ethic than the protection of privacy; therefore, it would be ethically justified to access the private information to facilitate the saving of a life.

Another case in point of this ethical worldview is the issue of using genetic profiling to validate the "worth" of a preborn individual. The use of genetic testing of human babies while they are in utero is becoming more commonplace. On one hand, this testing can detect health-related problems that can be corrected prior to birth. In other instances, test results are used to inform parents about health-related problems and associated counseling might suggest the abortion of the baby. In the latter case, the ethic of hierarchicalism could be applied, and the higher ethic would be saving the baby's life rather than the destruction of life.

Again, the impact of employing this worldview can be profound. As science and technology continue to advance, ethical conflicts will increase in both number and scope. All individuals will be placed in situations in which ethical decisions must be made with regard to technology; nobody will escape the philosophical mandates that all of humankind will face in some capacity or other. The only question that remains to be answered is what ethical worldview will be employed to address these situations? Each worldview provides an explanation of how to understand and help solve ethical problems. Technology educators have a responsibility to address these types of issues as part of the overall goal of encouraging technological literacy in the citizens of planet Earth.

## ROLE OF TECHNOLOGY TEACHER EDUCATION IN THE TEACHING OF ETHICS

During the fall of 1991, a group of recognized experts from the field of technology education met together to identify the essential framework components for the secondary level curriculum in technology education (Wicklein, 1992). The results of this project yielded nine primary curriculum areas that were considered necessary to create a successful instructional program for the middle school and high school technology teacher. As the principal researcher on this project, I found it very interesting that the primary curriculum goal identified by the panel of experts was "developing human potential" (p. 24). This curriculum goal addressed topics such as enhancing students' positive self-image, developing appropriate social skills, encouraging and developing student leadership skills, as well as others that had nothing to do with technology per se. The experts in this project defended this goal and the priority that it held in the curriculum framework by explaining that the most important role of any teacher is to help their students develop as responsible humans first and foremost, apart from teaching any particular, specific subject matter. With this as a base component in the secondary curriculum for technology education, it is clear that instruction on ethics is needed and appropriate for inclusion at all levels of instruction.

Common practices in technology teacher education do not attend to a detailed study and dialogue on ethical issues. Technology teacher educators often portray an ethical worldview of *materialism*. When this worldview is applied to the study of technology, and specifically to the teaching of technology, the focus is exclusively on the mechanisms of a given technological topic. Instruction and learning often focus on how technological things work (for example, building a model rocket, assembling an electrical circuit, creating a video advertisement). The ultimate purpose of this type of instruction is to inform and train students about specific technologies. Thoughtful considerations related to the values of the technology, impacts of the technology on society, and philosophical discussions on the significance of a given technology are often neglected or addressed in a most brief way. Teacher educators of technology education should examine their instructional content to ascertain whether ethical issues associated with technological systems that are covered in their classes are addressed.

A missing component in some technology teacher education curricula is *perspective*. Perspective, in this case, indicates the need to examine, not just where we are and where we are going with regard to technology, but what ought we think and do about technology. With current curricular approaches in technology education, students will emerge with a lopsided view of reality if educators do not address the entire progression of technology—past, present, and future—as well as the ethics and morals of technology.

The question of what to include when studying technology or any school subject is often critical for teachers, but according to Neil Postman (1992), author of *Technopoly: The Surrender of Culture to Technology*, this is of little importance. "Perhaps the most important contribution schools can make to the education of our youth is to give them a sense of coherence in their studies, a sense of purpose, meaning, and interconnectedness in what they learn" (pp. 185–186). Postman continues,

> Modern secular education is failing not because it doesn't teach who Ginger Rogers, Norman Mailer, and a thousand other people are [as well as snap grids on CAD, Flash video downloads, and CNC codes] but because it has no moral, social, or intellectual center. There is no set of ideas or attitudes that permeates all parts of the curriculum. The curriculum is not, in fact, a "course of study" at all but a meaningless hodgepodge of subjects. It does not even put forward a clear vision of what constitutes an educated person, unless it is a person who possesses "skills." In other words, a technocrat's ideal—a person with no

commitment and no point of view but with plenty of marketable skills. (p. 186)

Therefore, the question remains, what is the role of the technology teacher educator in the instruction of ethics? How does the technology teacher educator go about addressing the impacts of technology on society with ethics in mind? We cannot and should not expect the average classroom teacher of technology to tackle this topic on his or her own. The subject of ethics is a difficult one all by itself, but is compounded exponentially when placed within the politically charged climate of the public school arena. Technology teacher educators need to develop model instructional strategies wherein ethical issues are discussed and demonstrated for preservice teachers. Positive examples are needed to encourage the next generation of technology teachers to tackle the "prickly pear" of ethical instruction in the classroom. Ethical issues related to the development, use, and/or elimination of technology is important and essential to a comprehensive program of study in technology education. Teacher educators have a unique opportunity to lead the way on this important topic. Without our thought and example, we cannot assume that this type of instruction will ever be dealt with in an inclusive way. Ethics instruction is a challenging topic in our field and it can be a great opportunity to expand our horizons to consider all of the ramifications of the impact of technology on society, both here in the United States and around the world. The challenge is real—the question is, will we tackle it in a way that would do justice to a study of ethics?

When considering the impact that technology has on society, ethics is a central construct for this topic. It would seem very natural to consider the ethical ramifications of various technological systems if one could move beyond the mere technical features in the study of artifacts. For example, when studying energy systems, educators could easily move class activities and discussions to consider the relevance of alternative energy sources and compare these with more conventional energy production systems. In addition to comparing costs of production and output capabilities, teachers and students could debate the ethics of further development of alternative energy. Issues related to "ought" could and should be brought to the forefront of these types of instruction. Ethical reflection and deliberation should be viewed as not just another add-on feature to the study of technology but as a central and integral part of the core of understanding technology.

## INTERPRETING THE *STANDARDS FOR TECHNOLOGICAL LITERACY* WITH ETHICS INSTRUCTION

Ethical issues have been addressed and interwoven in several of the *Standards for Technological Literacy*. In particular, Standards 4, 5, and 6 speak to ethical concerns that are of great importance to our society. Standard 4 states, "Students will develop an understanding of the cultural, social, economic, and political effects of technology" (International Technology Education Association, 2000, p. 57). Within this standard, specific ethical points are posed that are important to consider in the development of the technologically literate person. For example, when discussing the changes brought to society by technological development, this standard presents the following ethical scenario: "Traditional ways of life have been displaced by technological development. This trend tends to magnify the inequalities among peoples and among societies by creating a situation in which a minority of people and groups control and use a majority of the world's resources" (p. 57).

Teacher educators can incorporate issues related to Standard 4 within their teacher preparation programs by providing real-world ethical situations and scenarios that typify this ethical problem. For example, it is a fact that U.S. companies own or control over 50 percent of the wealth that exists in the world today, and at the same time, the United States represents approximately 5 percent of the total population on Earth (Whaley, 1987). Through this simple statement of distribution of wealth, questions can be posed that challenge students to consider the ethics of this condition. These questions might include the following:

1. Is the distribution of the world's wealth just?
2. Should it be changed?
3. If so, how might you change it?
4. Have you ever experienced a similar situation in which something was distributed so unevenly? What did you do?
5. What role does technology play in the distribution of wealth?

When teacher educators integrate ethics instruction within their technology courses, preservice teachers begin to recognize the need and importance of these types of issues and learn how to incorporate ethical issues within their future classroom instruction.

Several of the standards' benchmarks focus specifically on ethics; for example, Standard 4, benchmark E states, "Technology, by itself, is neither good nor bad, but decisions about the use of products and systems can result in desirable or undesirable consequences" (International Technology Education Association, 2000, p. 60). This very statement could be questioned from an ethical perspective, as well as debated by considering a variety of technologies. Benchmark F of Standard 4 speaks specifically about ethics when it states "The development and use of technology poses ethical issues" (p. 61). In addition, benchmark J of Standard 4 addresses ethical matters by stating "Ethical considerations are important in the development, selection, and use of technologies" (p. 63). Other benchmarks, although not mentioning ethics within their description, bring into focus ethical dilemmas that are of great importance as students grow in their technological literacy. Benchmark K of Standard 4 states, "The transfer of a technology from one society to another can cause cultural, social, economic, and political changes affecting both societies to varying degrees" (p. 63).

As students consider how technology transfer impacts people, they should be encouraged to consider ethical problems that are brought to bear in these circumstances. For example, when Intel Corporation decided to move one of its microchip production facilities to San Jose, Costa Rica, they probably based the bulk of their decision on economic reasons (for example, the cost of labor in Costa Rica is much lower than in the United States). However, another important issue also needed to be considered; that is, what would be the social and cultural impact of building and running a production plant in Costa Rica? Is it ethical to disrupt the social and cultural structure of this country by tempting people to abandon their traditional social and cultural structure to take advantage of the jobs created by Intel? This is what benchmark K of Standard 4 is alluding to.

Standard 5, "Students will develop an understanding of the effects of technology on the environment" (International Technology Education Association, 2000, p. 65) and Standard 6, "Students will develop an understanding of the role of society in the development and use of technology" (p. 73) both highlight topics in which ethics is of special concern and interest. Teacher educators can easily identify examples within each of these two standards in which ethical questions surface and provide important teaching/learning moments. For example, within Standard 5 on *understanding the effects of technology on the environment*, ethical discussions and questions can be presented to address the concept of recycling. Are the current technological practices that support recycling of materials providing a positive environmental result, or are many of the current recycling processes merely a form of downcycling, which has more of a negative impact on the environment than the processing of virgin materials? According to international environmental architects McDonough and Braungart (2002), the current recycling practices often actually cause more environmental degradation and biosphere contamination. Debating the impacts and values of recycling would be a powerful learning experience as technology teachers and students reflect on the ethical ramifications of this practice that is typically assumed to be friendly to the environment.

Likewise, when addressing Standard 6 on *understanding the role of society in the development and use of technology,* students can examine ethical factors that lead societies to adopt or not adopt specific technological systems and processes. For example, students can carefully evaluate the rationale that certain groups of people have given for not using and adopting advanced technological systems (for example, Amish communities in Pennsylvania). Through a careful and fair analysis of the reasons why these people do not adopt advanced technologies, students can begin to understand that the ethics within a society can impact technology and vice versa.

In several of the *Standards for Technological Literacy,* ethics is interwoven as an integral component in the development, use, and impact that technology has on individuals, society, and culture. Technology teacher educators must include ethics discussion and instruction within their teacher preparation programs to provide a complete perspective in the study of technology. If ethics instruction is not included in teacher preparation, the future of technology education will be anemic and shortsighted in the preparation of a technologically literate citizenry.

## INSTRUCTIONAL ACTIVITY

### Description

A classical debate is a way to address the issues of ethics and the assessment of technological impacts on society. As an instructional activity, students are required to consider the ethics of a variety of technological topics, all of which have impacts on society. The following lesson plan describes this assignment in a format that can be used in any technology education program.

## Assignment

Write two essays (one per day). Read the following questions and form an opinion about the issue. Write a five-paragraph essay about your opinion. Be sure that your essay reflects a specific opinion and does <u>not</u> explain both the pros and the cons of the issue. You should write why you agree or disagree with the statement, but not both. Each day, you will turn in one essay. Vary your topics; do not choose more than one topic from each area.

<u>The essays must be completed by the end of the week. If you are not in class, you must complete this assignment at home.</u> We will be discussing the responses next week and our next unit will center on your answers.

Remember—there are no wrong opinions if you can justify your reasons!

## Debate Topics

### Bio-Technology

1. Should animals be used for medical research?
2. Should embryos be cloned?
3. Should euthanasia (mercy killing) be legal?
4. Should prescription drugs be sold on the Internet?

### Communication

1. Should students be allowed to carry cell phones to school?
2. Should the development of Web pages be monitored to restrict information available on the Internet?
3. Should there be regulations concerning the use of cell phones in automobiles?

### Energy, Power, and Transportation

1. Should personal transportation laws (seatbelts, helmet, and so on) be required by law or be a personal choice?
2. Should parents be able to revoke a teenager's driver's license?

3. Should gas prices be regulated?
4. Is too much money spent on space exploration?

### Production

1. Should the United States change to the metric system?
2. Should children under the age of 16 be allowed to work in factories?
3. Should road construction be limited to nighttime in high-traffic areas?

## Requirements

Your five-paragraph essays must include:

- An introductory paragraph
- Three body paragraphs, each explaining a different reason for your opinion
- A conclusion paragraph
- A minimum of three sentences in each paragraph

## Evaluation

Each essay will be graded using the following grading scale:

- Are there five paragraphs with at least three sentences each? *30 points*
- Does the introductory paragraph provide an overview of the topic and state your opinion? *10 points*
- Do the three body paragraphs provide three different reasons to support your opinion? *45 points*
- Does the conclusion paragraph summarize your reasons for your opinion? *10 points*
- Is the paper free of grammatical and spelling errors? *5 points*

Following the essay assignment, debate topics are selected from the highest rated essay assignments. The class is then divided into groups of two or three students; one group of two or three students debates a technological topic with another group of two or three students. The following debate procedures are used to guide the process and teach students the classical method of debating.

## *Ethics Debate Procedure*

- Opening statement by affirmative (2 minutes maximum time)
- Opening statement by opposition (2 minutes maximum time)
- Argument by affirmative (6–10 minutes)
- Cross-examination of affirmative (5 minutes maximum)
- Argument by opposition (6–10 minutes)
- Cross-examination of opposition (5 minutes maximum)
- Closing statement by opposition (2 minutes maximum)
- Closing statement by affirmative (2 minutes maximum)

## *Materials Developed*

The debate unit should require each student group to develop a number of documents and records that support their position on the topic being debated. The group should use these documents and records during the debating process. The required debating documentation includes:

**Cover Page**

- Title
- Names of group members
- Class period
- Date
- Debate graphic

**Opening Statement**

- Must be typed
- Must be read, taking no more than 1 minute in length

**PowerPoint Presentation**

- Minimum of 15 slides
- Must use pictures and graphics, not from clip art
- Must be 6–10 minutes in length
- Printed in note format, six per page

*Ethics and the Assessment of Technological Impacts on Society*

**Outline of Presentation**

- Typed
- Directly follow the presentation
- Summary of topic
- May be used when delivering presentation
- One to two pages in length

**Questions**

- Ten questions that could be used on opposing position
- Ten questions that may be asked of your position

**References**

- Minimum of five resources
- Follow MLA format style

**Binding**

- All documents and records should be bound.
- The bound document should be aesthetically pleasing and neat.

Following the completion of these materials, the debates can begin with students following all procedures.

*Debate Evaluations*

Evaluation will be part of the debating process. A detailed evaluation should be conducted to provide students with feedback for self-improvement. In this case, a combination of peer evaluations and instructor evaluation is helpful for individual students to gain confidence in communicating ideas and concepts. By providing an averaged score of all class peers (those students in the class observing the two debate teams) with the instructor's evaluation, the student debaters can learn ways to improve their research and communication skills. The following rubric can be utilized to evaluate the debating process.

This instructional activity could help students in technology education programs to observe and understand the subtleties of the ethical concerns involved in a variety of technological applications. As students become more aware of these issues, they will then be more sensitive to the impacts that technology places on society and vice versa.

Table 4-1. Ethics Debate Evaluation Rubric

Topic: _____

Group members: _____

### Materials Turned In

| Cover Page | 5 | |
|---|---|---|
| Include topic? | 1 | |
| Include members' names? | 1 | |
| Include a graphic? | 1 | |
| Creative | 2 | |

| Opening Statement | 5 | |
|---|---|---|
| Typed? | 1 | |
| 10–12 point font? | 1 | |
| Proper length? | 1 | |
| Quality | 2 | |

| PowerPoint Presentation | 15 | |
|---|---|---|
| Printed 6 slides per page? | 2 | |
| At least 15 slides? | 4 | |
| Relevant researched material? | 5 | |
| Appropriate pictures/graphics? | 4 | |

| Summary | 10 | |
|---|---|---|
| Typed? | 2 | |
| 10–12 point font? | 2 | |
| 1–2 pages long? | 2 | |
| Adequately summarizes information? | 4 | |

| Questions | 5 | |
|---|---|---|
| Typed? 10–12 point font? | 1 | |
| Questions to be asked? | 2 | |
| Questions to ask? | 2 | |

| Closing Statement | 5 | |
|---|---|---|
| Typed? | 1 | |
| 10–12 point font? | 1 | |
| Proper length? | 1 | |
| Quality | 2 | |

| Resources | 5 | |
|---|---|---|
| Typed? 10–12 point? | 1 | |
| At least 2? Quality of sources? | 2 | |
| MLA style formatting? | 2 | |

| Materials Grade | 100 | |
|---|---|---|

### Actual Debate

| Opening Statement | 10 | |
|---|---|---|
| At least 1 minute? | 2 | |
| Quality of information | 4 | |
| Quality of presentation | 4 | |

| Presentation | 20 | |
|---|---|---|
| 6–10 minutes in length | 5 | |
| Follow script? | 5 | |
| Quality of information | 5 | |
| Quality of presentation | 5 | |

| Cross-exam—asking | 5 | |
|---|---|---|
| Use at least 10 questions? | 2 | |
| Did all group members participate? | 1 | |
| Based on group presentation? | 2 | |

| Cross-exam—answering | 5 | |
|---|---|---|
| Adequately answer all questions? | 4 | |
| Did all group members participate? | 1 | |

| Closing Statement | 10 | |
|---|---|---|
| Adequately summarize the topic? | 8 | |
| Is it presented professionally? | 2 | |

| Bonus | 10 | |
|---|---|---|
| Are all group members dressed professionally? | 10 | |

| Debate Grade | 100 | |
|---|---|---|

| Comments |
|---|
| |

This debate assignment was originally created and field-tested by Ms. Angela Hughes, technology teacher at Morrow High School in Clayton County, Georgia (Powell, 2002).

## REFLECTION QUESTIONS

1. Of the five worldviews described in this chapter, which one is most consistent with your ethical perspective?
2. How do your ethical beliefs impact your uses of technology?
3. If you were asked by an employer to apply technology to the solution of a problem in a way that violated your ethical standards (damaging the environment, creating a hazard for other employees, taking advantage of lower income workers, and so on), how would you respond?
4. As a technology teacher, how would you compare the importance of teaching students technical content with teaching them about the ethical ramifications of related technological systems?

# REFERENCES

Annas, G. J., Glantz, L. H., & Roche, P. A. (1995). Drafting the genetic privacy act. *Journal of Law, Medicine & Ethics, 23*, 360-366.

Brandt, R. B. (1967). Ethical relativism. In *The Encyclopedia of Philosophy*, (Vol. 3, pp. 75-78). New York: The Encyclopedia of Philosophy.

Campbell, K. (1967). Materialism. In *The Encyclopedia of Philosophy*, (Vol. 5, pp. 179-188). New York: The Encyclopedia of Philosophy.

Geisler, N. L. (1971). *Ethics: Alternatives and issues.* Grand Rapids, MI: Zondervan.

Kidd, I. G. (1967). Cynics. In *The Encyclopedia of Philosophy*, (Vol. 2, pp. 284-285). New York: The Encyclopedia of Philosophy.

Kreeft, P. (1996). *The journey: A spiritual roadmap for modern pilgrims.* Downers Grove, IL: Inter Varsity Press.

International Technology Education Association. (2000). *Standards for technological literacy: Content for the study of technology.* Washington, DC: National Science Foundation.

McDonough, W., & Braungart, M. (2002). *Cradle to cradle: Remaking the ways we make things.* New York: North Point Press.

Norman, D. A. (1993). *Things that make us smart.* Reading, MA: Addison-Wesley.

Popkin, R. H. (1967). Skepticism. In *The Encyclopedia of Philosophy*, (Vol. 7, pp. 449-461). New York: The Encyclopedia of Philosophy.

Postman, N. (1992). *Technopoly: The surrender of culture to technology.* New York: Vintage Books.

Powell, A. (2002). Ethical debates: Instructional activity for technology education. Unpublished manuscript.

Shrader-Frechette, K., & Westra, L. (1997). Overview: Ethical studies about technology. In K. Shrader-Frechette & L. Westra (Eds.), *Technology and values* (pp. 3-10). Lanham, MD: Rowman & Littlefield.

Vanderburg, W. H. (2000). *The labyrinth of technology.* Toronto: University of Toronto Press.

Whaley, L. E. (1987). *Future studies: Personal and global possibilities.* Monroe, NY: Trillium Press.

Wicklein, R. C. (1992). Curriculum development in technology education. *The Technology Teacher, 51*(5), 23-25.

# Developmental and Contextual Issues Related to Ethics and Character

## Chapter 5

Rodney L. Custer
Illinois State University
Normal, IL
&
Danny C. Brown
Illinois State University
Normal, IL

On September 29, 1767, a young Gambian man named Kunta Kinte arrived in Annapolis, Maryland, aboard the cargo ship, Lord Ligonier. This event would have amounted to little more than an obscure historical footnote were it not for author Alex Haley and his best-selling book *Roots*. The compelling story of a family's genealogy from slave ship through life in the rural south to professional life in America resonated with millions. Estimates are that, by the end of the 1977 miniseries, three out of every four American households with televisions were tuned into the program (The Beginning, n.d.).

Perhaps as much as any single factor, Haley's *Roots* is responsible for triggering a tremendous surge in interest in family genealogy. The dramatic growth of Web-based technology and the proliferation of powerful search engines and genealogical software programs have sustained the interest. Huge amounts of information about one's ancestors are now available. The mapping of the human genome and other information-intensive technological developments are almost certain to make the search for roots more accurate, compelling, and even perhaps disturbing.

The focus of this chapter is on the development of ethical and cultural genealogies. Initially, the issues appear to be rather clear. Ethical and cultural values are, for the most part, understood by the people who live in various countries and communities around the world. Most people generally understand what is considered to be acceptable and unacceptable behavior. Whether they choose to comply or not is another matter. Good behavior is rewarded by such things as community recognition, a sense of well-being, and an acknowledgment of membership or "fit" within the culture or community. "Bad behavior," depending on the severity, is punished formally by legal action or informally by such things as ridicule, gossip, and peer pressure. Social institutions such as the legal, religious, and educational institutions serve to codify and propagate cultural values. At one level, that's pretty much it. Cultural value systems tend to function quietly in the background, woven throughout the general context within which people live their lives.

Just as *Roots* presented the genealogy of a family, this chapter provides an examination of the genealogy (forces, events, institutions, and so on) that "gives rise" to cultural and social values and character. In the initial section of the chapter, we describe and elaborate on a conceptual framework designed to examine cultural and ethical values within the larger context of human development. A key part of this discussion consists of thinking about how technology, in various ways, influences this process. We then move on to a discussion of two broad and interesting arenas in which culture, values, and technology interact. These are (*a*) consumption and consumerism and (*b*) power and authority.

## HUMAN DEVELOPMENT: A CONCEPTUAL FRAMEWORK

We begin the discussion with the introduction of a conceptual framework, developed by Welzel, Inglehart, and Klingemann (2003), dealing with social progress and human development. Essentially, the framework consists of three primary dimensions that form a coherent structure within which individuals and societies evolve and change. These dimensions are economic, cultural, and institutional (see Figure 5-1).

### *The Economic Dimension*

The economic dimension has to do with socioeconomic development and comprises a variety of basic individual needs elements, such as material prosperity, access to education, and socioeconomic status, as well as larger social processes, such as urbanization, social mobilization, and occupational differentiation. This dimension is the most fundamental motivator of social development and ethics. It is also generative, serving as an engine for social and cultural progress. As economic resources are developed and made available increasingly to individuals and communities, progress occurs (at least by measures that are typically associated with quality-of-life issues). The implications parallel those that Maslow made at the individual level, at which self-actualization tasks come into focus as

Figure 5-1. Human development model.

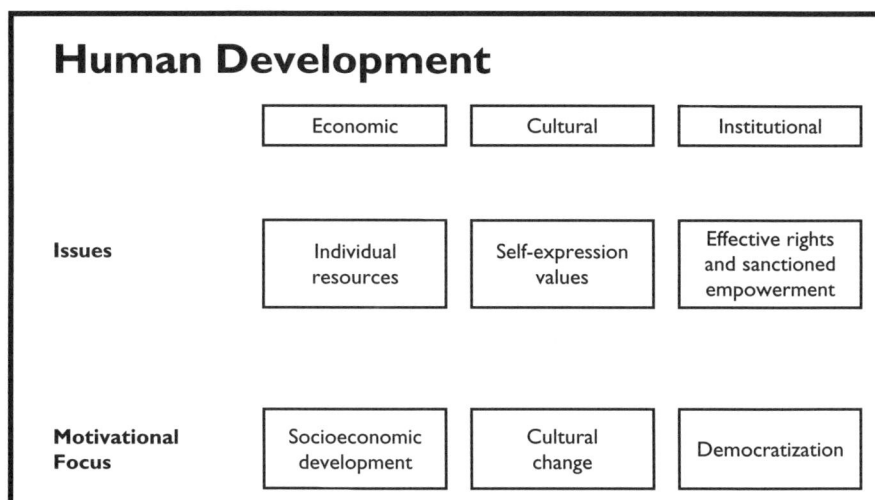

basic survival needs are met (DeCarvalho, 1991). Much the same occurs with societies and nations. Economic resources, particularly when they are in abundance and when they are distributed across society, have a positive and generating impact on how societies develop. This affects a range of things from governmental structures, cultural values, information distribution, and much more.

*The Cultural Dimension*

The cultural dimension focuses on cultural values and change. Here, the emphasis is primarily on self-expression, including such things as tolerance for diversity, civic involvement, concern for individual rights and freedom of choice, and security and trust of others. Within the larger context of human development, changes in the cultural dimension are heavily influenced by what happens in the economic dimension. When times are good and resources are available and well-distributed, values such as self-expression, freedom, and choice tend to emerge as well. According to

Welzel et al. (2003), growing individual resources widen the scope of human activities and heighten the level of possible achievements. The striving for self-expression finds greater leverage, which fuels growing public emphasis on human choice.

This is, of course, consistent with Western notions of progress, in which economic growth is closely aligned with social progress. Perhaps more than any other single idea, this fundamental notion of (and belief in) progress has affected how we view the cultural and ethical elements of society. Nisbett (1980, p. 8) noted that "religion, science, reason, freedom, equality, justice, philosophy, the arts, and so on" are all grounded in and influenced by the idea of progress. Thus, the idea of progress provides an ideological link between economic growth and cultural development. (It is important to note briefly that valid concerns have been and should be raised about this strong connection between progress and economic-social-cultural values. More of that discussion is presented later in the chapter.)

## *The Institutional Dimension*

The institutional dimension is concerned with political structures, particularly the ways in which economic growth and cultural development affect political structures. Within Welzel et al.'s (2003) human development model and from their Western perspective, economic growth tends to spawn a growth in self-expression values that, in turn, leads to an increase in democratic values and forms of government. Of course, the reverse can also be true, in which large-scale economic deprivation tends to repress individual choice and expression, leading to repressive and autocratic governmental structures.

Although this framework clearly reflects a strong Western bias, it, nevertheless, provides a useful conceptual structure within which to conduct our discussion of technology and ethics for several reasons. First, it provides a useful context that encompasses individual, social, cultural, and political spheres. Second, the framework spans a range of social institutions, including economics, politics, religion, education, and the family. Third, the framework provides a structure for exploring vast sociocultural differences. Finally, the framework provides a conceptual framework within which to examine technology within the larger context of human development.

## Technology and Human Development

Up to this point in our discussion of the human development model, the discussion has centered on economic, social, and political issues. When technology is factored into this model, some interesting observations can be made.

Clearly, technology is a major driver of the economic dimension. Technology has had an enormous impact on the efficient production of goods and services and is one of the most important factors affecting economic development. Technology and the generation of wealth simply go hand in hand.

On the institutional end of the scale, the picture is far less optimistic. As technology becomes increasingly more complex, there is a sense in which democracy is being eroded. Democracy, at its foundations, depends on an informed and participating citizenry. Increasingly, the average person is not well-equipped to make technological choices, and the overall level of technological literacy is eroding. Most decisions about technology are being made by persons who possess specialized expertise and knowledge. How much does the average person understand about such things as the internal workings of automobiles, the facilities that sanitize the water supply, how electricity is generated and distributed, and how products are designed and made? These are complex technological systems requiring specialized engineering knowledge to develop and maintain. Quite simply, the complexity of modern technology tends to work against and discourage citizen participation. Many of the most important decisions are being made in the absence of citizen participation and awareness. Technology has simply become too complex for the average citizen to understand. Given the importance of citizen participation for democracy, this basic lack of technological literacy should be a matter of fundamental concern.

If it is true, as Welzel et al. (2003) indicate, that the three dimensions presented in the model function as a coherent whole and if our analysis of technology's role in the economic and institutional dimensions is correct, a question must be asked about the implications for cultural change. What occurs in a society when economic issues have largely been *addressed* (through the use of technology) and, at the same time, democratic foundations are, at least to some extent, being eroded by *technology*?

We posit that the dynamics created by technology (both positive and negative) have some powerful implications for values and ethics. The seeds are there for powerlessness, apathy, passivity, cynicism, and dependence rather than active participation, engagement, creative expression, and positive influence. When technology is factored into the human development model in highly developed democratic cultures, we argue that the prevailing causal direction has been reversed (see the dotted arrow in Figure 5-1). Unlike the situation for developing countries and emerging democracies, in which economic growth spawns values such as increased openness, self-expression, cultural richness, valuation of freedom, and democracy, we argue that, in powerful and established democratic cultures, technology often serves to erode culture.

Consider, for example, the homogenization of America (and the world) in which increasingly every community is coming to look like every other. Local hardware stores, locally owned grocery stores, the gas station with a mechanic, and a variety of small shops have given way to the Wal-Marts, Home Depots, Domino's Pizzas, McDonald's and any number of other large national chains. Although there is nothing inherently bad about large chain stores, it is clear that the drive for economic efficiency coupled with powerful information management technologies have combined to flood the landscape with stores and businesses that can be easily replicated across the country and world.

If our analysis is true, there are serious implications for the cultural dimension of the human development model. Homogenization, by definition, leads to less diversity and self-expression. When the design and production of goods and services are increasingly being relegated to professionals (and their sophisticated machines), what are the implications for individual creativity, ingenuity, and self-expression?

Up to this point, we have been discussing the larger human development context and framework within which ethical decisions are made. We have also explored some implications of superimposing technology onto the model and have raised concerns about the status of culture in highly developed, democratic societies. The focus now shifts to two of the most critically important areas in which technology, social values, and ethics interface with one another—namely, consumption and power.

## TECHNOLOGY AND CONSUMPTION

In 1998, Dwight Murphey published an article titled *A Revolution in Progress: Today's Proliferating Science and Technology*. The article described a number of fascinating and futuristic technologies that were expected to emerge in future years. Among these were such things as artificial intelligence and expert systems in manufacturing, telemedicine, nanotechnologies, speech recognition, virtual offices, electronic money, and much more. The article was striking in several respects. First, it gave evidence of technological change that is continuing to accelerate from its already rapid pace. Second, technology is extending its reach into virtually every aspect of human life, including the development and production of agricultural and manufactured products, the ways in which the corporate world processes information, and the emergence of sophisticated biotechnologies. Increasingly, technology is being woven into the very fiber of human existence.

The most striking observation, however, to be made from Murphey's article is that almost everything that was projected to occur in the development of technology just six years ago has (at least to some extent) already occurred! We are indeed, as Jencks (1986) observes, living in a postmodern world characterized by such things as postindustrialism, the information revolution, consumerism, multinational economics, high-speed communications, immigration, pluralism, eclecticism, inclusivism, and multiculturalism.

The pace of technological change could lead to the conclusion that the sky is the limit when it comes to the possibilities for new technological development. To some extent, this is true. Any reasonable analysis of recent decades would lead to the expectation that future technological development will yield astonishing biomedical advances, better products, faster and better communications, and so on.

However, when the focus shifts to the moral and ethical dimensions of consumption, the issues become more complex and troublesome. Borgmann (2000), in a thoughtful critique of contemporary culture, observed that "the moral condition that typical human condition inspires today is not outrage or indignation but the sort of searing regret one feels when something beautiful is being defaced by neglect" (p. 192). Citing the remarkable engineering challenges and ingenuity leading to the development of high-definition television as an example, Borgmann goes on to observe that "We watch on television what and where we would like to be, outside somewhere, bravely and skillfully facing real challenges, but we never get around to doing and being what we watch. . . . The ingenuity of construction will yield to banality of consumption" (p. 191). It is symptomatic that the leading causes of death in the United States have to do with consumption and lifestyle issues (diet, smoking, lack of exercise, and so on) and include strokes, heart disease, cancer, and diabetes.

Thus, it is clear that although technology is key to the advancement of the human condition, important moral and ethical concerns must be addressed as well. In technologically advanced, democratic cultures, many of these ethical issues have to do with consumption (overconsumption) and the use of the world's resources. We identify and discuss several key ethical issues that focus on the relationship between consumption and technology.

## HAPPINESS AND QUALITY OF LIFE

The concept, quality of life, is a complex one and is defined differently by various groups. Humanitarian groups such as the *Bread for the World Institute on Hunger & Development* use criteria such as infant mortality, female illiteracy, household access to running water, employment, and access to health care (Chelliah, 1992) to evaluate quality of life. By contrast, the indicator of choice for many is economic. In tangible form, quality-of-life indicators for people in affluent Western societies translate into a dizzying array of consumer goods, including automobiles, homes, entertainment equipment, electronic gadgets, foodstuffs, and more.

Psychologists have generally found that wealth is a poor indicator of happiness. "People have not become happier as their societies have become richer" (Eckersley, 1999, p. 24). Chilean economist Manfred Max-Neef, in a survey of 19 countries (both rich and poor), reached a similar conclusion that "societies experience a period in which economic growth brings about an improvement in quality of life, but only up to a point—the threshold point—beyond which more economic growth may lead to a deterioration in quality of life" (cited in Eckersley, p. 25).

From a moral perspective, it is important to ask, "What is the potential (and what are the limits) of technology and the things we consume when it comes to making people happy and enriching existence? What measures, in addition to economic, are appropriate for defining quality of life? How can people be encouraged appropriately to broaden their perspectives on consumption and quality-of-life issues?"

## RESPONSIBLE CONSUMPTION

In an insightful discussion of technology in America, Marcus and Segal (1999) described how emphasis shifts back and forth between individual and social issues. The Green Revolution of the 1960s and much of post-WWII technological achievements, including the development of nuclear power and space exploration, were focused at a level beyond the individual. The post-Watergate, Vietnam, civil rights era saw the transformation of complex social issues into concerns for individual rights and freedoms.

This dichotomy has important implications for the economics and ethics of consumption. Alcorn (2003) identified five factors that collectively influence economic growth: population, state of the arts, growth of knowledge, available resources, and rate of capitalization. Two of these factors, the growth of knowledge and state of the arts (techniques designed to improve efficiency, quality, and so on), are essentially technical issues directly related to technological capability and generally involving the work of the engineering and scientific communities. The other issues, which are political and ethical, stand at the juxtaposition of the individual and social good. Policies related to population growth and control are among the most serious issues facing governments and policy makers on a global scale. The rate of capitalization has to do with the extent to which society is willing to invest in economic growth. Decisions at this level directly impact individual social and economic well-being and often run counter to what may be in the best economic interests of the country. Issues here include such things as monetary policy, taxation, balance of trade, and investment in infrastructure.

Distribution of resources is perhaps the most problematic of all of the factors related to economic growth. The world's resources are finite and they are being consumed disproportionately by a relatively small proportion of the world's population. One-half of Africa's 600 million people are living on less than one dollar per day (Amin and Dubois, 1999). One of the most important (and potentially destabilizing) ethical issues that must be addressed in the twenty-first century focuses on how we balance the "good" for the individuals with what is in the best interest of the larger society and global community. How do we appropriately balance individual opportunity, which is fundamental to capitalism, with concerns for equitability, poverty, and justice on a global scale?

At this point, it is helpful to comment briefly on how cultural values can be viewed appropriately from at least two perspectives: individual and social. From the larger social perspective, the focus shifts to a range of larger social-cultural-political issues, including the role of social institutions and cross-cultural assimilation, change, and conflict. From the individual perspective, the issues have to do with such things as psychosocial development, family and community influences, and local cultural patterns. Although the distinction between these two is somewhat contrived, the developmental issues associated with each tend to be viewed differently. So, for the purposes of this discussion, we have decided to draw the distinction.

At first blush, it appears that the social level represents an extension or scaling up of individual needs and concerns. When viewed from the perspective of consumption, however, individual and social needs often stand in direct conflict with one another. Given the certainty of ongoing technological development, the finiteness of the world's resources, and the growth of communications technologies, which are exposing the world's poor to what others have, ethical issues related to consumption are among the most important issues of the twenty-first century.

## MANAGING POWER AND AUTHORITY

The issues of power and authority discussed here encompass not only instances of authority exercised over others as practiced by governments and hierarchal workplace structures, but also individual issues, such as the effects of technological changes on self-esteem and locus of control. Individual concerns about the sense of disenfranchisement brought about by inequalities in access to opportunity are also considered. Discussion of issues of power and authority in this context demands exploration of belief systems and the conflicts that may arise from their differences, efforts to discover common ground in ethical positions, impacts of technology on democratic principles, and a brief examination of the role of technology in balancing power.

### *Belief Systems Sometimes Result in Conflict*

As previously mentioned, one characteristic of open societies is their tendency to move toward homogenization with other societies exposed to common influences. Language, accents, dress, and other once stark cultural differences diminish as persons from open societies around the world watch the same television shows, read similar books, listen to syndicated radio programs, and digest newspaper articles originating from common news bureaus.

In contrast, closed societies with rigid traditions more often define ethical behavior and limit reflection on the need for changing moral standards. Stark differences in perception can result in conflict between closed societies, often presented and enforced as inspired, self-evident, and unquestionable (Bronowski, 1956), and open societies that embrace changing and often pluralistic views of morality and ethical behavior. It has been suggested that this homogenization of culture, ethics, and values brought about by exposure to the openness of cyberspace-based communication may bring about a future of unprecedented conflict. This could be a future in which previously isolated, closed societies may increasingly feel threatened by the encroachment of more open, globalized societies. This threat is exacerbated by fear within the leadership of closed societies that open; uncontrolled communication may ultimately encourage movement toward demands for greater democratization (Barlow cited in Regan, 2001).

It has been suggested that most, if not all, wars have originated not in the scarcity of resources, but rather from the inequitable distribution of the resources that are available as well as some person's desire to control even more resources than were previously controlled (Fuller, 1981). Issues of control and access to resources and the wealth they represent are constantly redefined by the advent of new technologies, which in turn create shifts in wealth through introduction of new uses and demands for limited resources. The widespread dependence on the automobile has created a shift in moral perspective, which culminated in widespread citizen support for sending U.S. troops into Kuwait in 1990 to expel Iraq. If vast oil reserves had not been present in Kuwait, it seems unlikely that the public would have considered this military action to be a moral imperative. Each shift in resource demand creates additional pressure on people to reflect upon the standards by which they view the world and their relationships with members of other groups.

Conflicts between belief systems can also emerge between individuals and the community. In the 1952 (Vonnegut) novel, the *Player Piano*, a world was described in which the technological ability to mass-produce enormous amounts of goods with almost no actual physical labor contributed to a world in which only a few technologically gifted people had opportunities to contribute meaningfully or creatively. This was a world in which the responsibility of the average person was to be a passive consumer of goods, thus providing a reason for the manufacturing machines to continue producing products. The work of most people was mindless service work created to keep people moderately busy and under control. The ethical standards of the time, dictated by the prevailing community to preserve the status quo, required people to accept this world without question. Demand for individuality or meaningful work was not allowed, and to question or withdraw from the system was to become an outlaw and a pariah. The toll this took on individual self-esteem and people's sense of locus of control was enormous. The price paid by the individual for living in such a dehumanizing world was terrifying. Although the scenario described in Vonnegut's work has thankfully not been realized, elements of this sort of conflict and its resulting isolation and powerlessness occur routinely. Global economic competition may create the need for ever higher levels of worker productivity, which brings near-term economic advantage but often at the long-term cost of greater individual dehumanization and alienation.

Perhaps the greatest defense for protecting and empowering citizens in a technological world is cultivated when technology educators help individuals develop a healthy sense of both empathy and skepticism. Skepticism and empathy must derive not only from individual reflection, but also through social dialogue, as values-based discussions that include a full range of economic, cultural, and institutional dimensions. Questions about how we ought to behave are social questions that must be defined (and redefined) socially to be meaningful (Bronowski, 1956). This is not to suggest that whatever value system is in place must be rejected and reinvented by each generation. Rather, we suggest that discussion of inequities in our value systems should be promoted. To learn to support the moral values of the community, children must experience opportunities to practice empathy and skepticism and to reflect on what the actions and characteristics of a good person ought to be. They need to explore how and why the values of their community came to be. They need to also reflect upon which traditions and values are worth preserving and learn how to address those that are no longer appropriate (Glazner, 2001; Kohn, 1997).

## Seeking Points of Agreement

Empathy (Etzioni, 1998; Goldman, 1993) and self discipline (Etzioni) have been proposed as the ideals of ethical thinking that should be taught in school. The Aspen Declaration was a 1992 conference of ethicists and educators who convened to initiate a set of common principles intended to encourage a movement to include universal character education in the curricula of public schools in the United States. The six basic universal values that this growing educational movement revolves around are trustworthiness, respect, responsibility, fairness, caring, and good citizenship (Josephson Institute, 2002).

The debate about teaching ethics in schools heats up when we get to the question of which values should be taught. As frequently discussed, character education, designed to teach moral and ethical behavior focused on concepts such as respect, responsibility, and citizenship, seems difficult to argue against, but sometimes these values are operationally defined very differently by various faiths, cultures, and economic systems. This problem of differences in definition is exaggerated if respect, responsibility, and citizenship become "euphemisms for uncritical deference to authority" as Kohn (1997) suggests has occurred in some implementations of character education curricula.

In addition to the work of those actively involved in the debates about and implementation of character education, one idea for preparing students to cope with the complexities of the technological world of the future has been described as a postmodern approach to Science Technology and Society (STS). It has been suggested that the postmodern approach to STS should go beyond traditional discussions of environmental and economic consequences to include analysis of ecological, moral, cultural, pluralistic, and spiritual perspectives when exploring emerging technologies. This approach would center on an empathetic ethic of caring and a critical pragmatism, as diverse peoples consider how to create a better world for themselves (May 1992).

## Democratic Principles

Two traits that characterize the democratic society are reliance on the recognition of mutual interests as a factor in social control and free interaction between social groups creating opportunities for continuous readjustment to new situations through dialogue (Dewey, 1944). In 1970, Muller observed that over the last 100 years, science and technology have moved from bodies of knowledge that could be mastered by the typical "country gentleman" to something so complex and prolific that even the specialist was likely to struggle to keep abreast of the latest information and discoveries. Few would argue against the belief that the rate of creation of new technology and scientific discovery has continued and even accelerated since the 1970s. Now, the lay observer typically finds it impossible to acquire sufficient expertise and understanding of complex technologies to allow meaningful input into important political decisions of a technological nature. This sense that there is little opportunity for meaningful input into the democratic process on technology-related issues tends to discourage involvement in participatory governance. Helping citizens develop a greater level of comfort with reflective discussion of moral dilemmas brought about by technological change may be one important way of keeping the democratic dialogue going.

## *Role of Technology in Balancing Power*

The advent of iron weapons allowed creation of weapons technically superior to those made from bronze. The widespread manufacture of the crossbow spelled doom for the mounted knight. The introduction of reliable, repeating firearms spelled an end to the concept of "civilized" warfare, which was designed to limit death and injury to noncombatants. Our degree of power over self, our environment, and others is constantly in a state of flux. Technology, in controlling the environment to better meet the wants and needs of persons, fosters realignment of power from one individual or group to another.

In recent decades, disparities in access to quality education and to acquisition of computer literacy have created a future likely to place even greater pressure on the democratic ideal of equality in opportunity. When such situations create a sense that control of opportunities and access has shifted from within the individual to some external force, alienation and a further sense of powerlessness are likely to develop. This state increases the probability of future conflict within our own nation.

## SUMMARY

We often speak of technology as value neutral, as the product of attempts to meet the wants and needs of humanity. Our assumption is that we create technology, and we control it and its application. What we too often overlook is that while we attempt to control technology, it also is controlling us, often in ways that threaten our very understanding of what is good and bad. Technology has created scenarios that challenge us to reconsider what we perceive to be sanity and madness, beautiful and ugly, just and unjust, rational and irrational (Grant, 1986). Technology has provided us with amazing opportunities. The eradication of once dread diseases such as polio and tuberculosis are now within the realm of possibility. The average life expectancy has risen dramatically. We have the technological ability to desalinate sea water to provide drinking water in regions in which water is scarce. We have the ability to transport food from locations where surpluses could be wasted to locations where drought threatens starvation. We have the ability to convert renewable energy sources to light our homes and power our vehicles. We have brought the cost of most goods down and raised the standard of material living in most developed countries.

The ethical standards we each accept are bound up with our sense of identity. They are inseparable from our place in a community, our traditions, our understanding of the world, and human nature. Earlier, we considered the application of the human development model based on the interaction of economic, cultural, and instructional components (Welzel et al., 2003) to help us examine the development of ethics and character. What our examination has suggested is that perhaps that model is more cyclical than linear and that it is impossible to separate the influences of technology from this cycle of development.

## REFLECTION QUESTIONS

1. How has technology eroded the opportunities for individuals to be involved in decision making about the world around them?
2. In what ways might advanced technological development limit opportunities for a democratic way of life in the United States?
3. What is the likelihood you would be happy if you were wealthy?
4. How would you feel if you had to survive on less than one dollar per day? Does the topic of ethics have anything to do with the fact that about 300 million people in Africa are living on less than one dollar per day?
5. What traditions and values in your community should be preserved in coming generations? What traditions and values would your community be better off without?
6. Historically, what has been the relationship between technology and power? Is there still a relationship between technology and power?
7. What arguments can you make that technology is not value neutral, but is either good or bad?
8. What points would you include in a rationale for collaboration between technology teachers and social studies teachers based on materials presented in this chapter?

# REFERENCES

Alcorn, P. A. (2003). *Social issues in technology: A format for investigation* (4th ed.). Upper Saddle River, NY: Prentice Hall.

Amin, A. A., & Dubois, J. L. (1999). Cameroon poverty profile: Reducing the current poverty and tempering the increase in inequality. Washington, DC: World Bank/IMF/Cameroon.

The beginning: The roots of the Kunta Kinte-Alex Haley foundation. (n.d.). Retrieved December 24, 2003, from http://www.kintehaley.org/beginning.html

Borgmann, A. (2000). Society in the postmodern era. *Washington Quarterly, 23*(1), 189-200.

Bronowski, J. (1956). *Science and human values.* New York: Harper Row.

Chelliah, D. (1992). Middle-east hunger update. In M. J. Cohen (Ed.), *Hunger 1993: Uprooted People* (pp. 148-153). Washington, DC: Bread for the World Institute on Hunger and Development.

DeCarvalho, R. J. (1991). *The growth hypothesis in psychology: The humanistic psychology of Abraham Maslow and Carl Rogers.* Lewiston, NY: Edwin Mellen Press.

Dewey, J. (1944). *Democracy and education.* New York: Free Press.

Eckersley, R. (1999). Is life really getting better? *The Futurist, 33*(1), 23-26.

Etzioni, A. (1998). How not to discuss character education. *Phi Delta Kappan, 79,* 446-448.

Fuller, R. B. (1981). *Critical path.* New York: St. Martin's Press.

Glazner, P. L. (2001). Exit interviews: Learning about character education from post-Soviet educators. *Phi Delta Kappan, 82*(9), 691-693.

Goldman, A. I. (1993). Ethics and cognitive science. *Ethics, 103,* 337-360.

Grant, G. (1986). *Technology and justice.* Notre Dame, IN: University of Notre Dame Press.

Jencks, C. (1986). What is postmodernism? *Oxford Paperback Encyclopedia.* Oxford University Press, p. 47.

Josephson Institute of Ethics. (2002). *Making ethical decisions.* Retrieved April 3, 2002, from http://www.josephsoninstitute.org/

Kohn, A. (1997). How not to teach values: A critical look at character education. *Phi Delta Kappan, 78,* 428-439.

Marcus, A. I., & Segal, H. P. (1999). *Technology in America: A brief history.* New York: Harcourt Brace.

May, W. T. (1992). What are the subjects of STS—really? *Theory into Practice, 31*(1), 73-83.

Muller, H. J. (1970). The children of Frankenstein. Bloomington: Indiana University Press.

Murphey, D. D. (1998). A revolution in progress: Today's proliferating science and technology. *The Journal of Social, Political and Economic Studies, 23*(1), 75-92.

Nisbett, R. (1980). *History of the idea of progress.* New York: Basic Books, Inc.

Regan, T. (2001). The true battlefront of the 21st century: Open systems versus closed systems. *Christian Science Monitor.* Retrieved July 5, 2002, from http://www.csmonitor.com/2001/1113/p25s2-stin.html

Vonnegut, K. (1952). *Player piano.* New York: Delacorte Press.

Welzel, C., Inglehart, R., & Klingemann, H. (2003). The theory of human development: A cross-cultural analysis. *The European Journal of Political Research, 42*(3), 341-379.

# The Status of Ethics in Technology Education

## Chapter 6

Philip A. Reed
Old Dominion University
Norfolk, VA

Angela Hughes
Morrow High School
Morrow, GA

Susan Presley
North Cobb High School
Kennesaw, GA

Diane Irwin Stephens
Jasper County Middle/High Schools
Monticello, GA

Ethics is not a new concept within technology education. The inclusion of ethics evolved naturally from the progression of technological activity in the latter part of the twentieth century. During this shift to a postindustrial society, people started to look at technology from a more humanistic view than they previously had. To keep pace with these changes, a "new ethic" was suggested to help advance technological literacy by highlighting the relationship between humans, the environment, and technology (DeVore, 1980, 1991).

How far have we come? This chapter reviews the current state of ethics within technology education. In the first two sections, materials for classroom instruction, including textbooks and modular materials, are examined. The third section discusses and recommends resources and practices that appear in professional literature. A survey of international technology education and ethics constitutes the fourth section. The chapter concludes with a look at professional ethics as they relate to technology teachers, teacher educators, and administrators.

## ETHICS MATERIALS IN TEXTBOOKS AND OTHER PRINTED MATTER

This section focuses on the incorporation of ethics in textbooks and other printed materials that are available for use in technology education programs. Five textbook publishers were invited to participate in a survey intended to identify ethics-related materials that are available within various components of technology education. The survey was not designed to critique textbook publisher materials. The companies were asked if they produced curriculum materials that contained instruction in ethics. Two textbook publishers, Goodheart-Willcox and Pearson Education, chose to participate in the study. Both of these publishers identified that they produced curriculum materials containing instruction in ethics, including textbooks and/or workbooks as well as teacher instructional materials. One of the two vendors also stated that in addition to the textbook and/or workbooks and teacher materials, they also produce student-centered activities and software/multimedia materials. Both vendors provide curriculum materials for grades 6–8, but one also generates instructional materials for high school students that incorporate ethics. In their responses, representatives from Goodheart-Willcox and Pearson Education stated that their curriculum materials are aligned with *Standards for Technological Literacy: Content for the Study of Technology* (International Technology Education Association [ITEA], 2000). In addition, one of the vendor's materials are aligned with the Texas Education Agency.

## *Topics Covered*

The 20 *Standards for Technological Literacy* highlight 17 areas that are important for students to study to become technologically literate. Participants of the survey were asked to identify which of the following categories their curriculum materials teach about ethics:

1. Characteristics and scope of technology
2. Core concepts of technology (systems, resources, requirements, optimization, and trade-offs)
3. Relationships among technologies and connections between technology and other fields
4. Cultural, social, economic, and political effects of technology
5. The effects of technology on the environment
6. The role of society in the development and use of technology
7. The influence of technology on history
8. Design
9. Problem solving
10. The impacts of products and systems
11. Medical
12. Agricultural and biotechnology
13. Energy and transportation

14. Information and communication
15. Transportation
16. Manufacturing
17. Construction

One of the textbook vendor respondents stated that their curriculum materials incorporate ethics with all 17 of the categories identified in the *Standards for Technological Literacy*. The other representative stated that their ethics materials incorporate all of the categories excluding the influence of technology on history and the impacts of products and systems.

*Presentation Strategies*

Publishing companies were asked to identify the presentation strategies utilized in their textbooks or other printed materials when incorporating ethics. In their surveys, both of the participating providers of textbooks and other printed materials stated that ethics-related instruction was presented in their materials through reading and writing activities as well as through teacher presentations; however, the intensity of the instruction varied between the two. One vendor identified the ethics-related curriculum materials as comprehensive, including several lessons or activities and, in addition to the reading, writing, and teacher activities, also used research, discussion, and debate as presentation strategies. The other vendor indicated that ethics was mentioned in the materials, but no specific lessons or activities for teaching ethics were included.

*Expected Outcomes*

The textbook publishers were asked to identify the strategies in which students were assessed. Both respondents indicated that assessment was performed through written work (excluding tests or quizzes) and student presentations. One of the companies also identified tests and quizzes as well as class or group discussions as strategies for assessment.

To determine the effectiveness of the curriculum incorporating ethics, both companies' materials were reviewed through instructor surveys prior to publication and after dissemination. One of the companies also utilized student surveys to review the effectiveness of the materials. In addition to being reviewed prior to publication, one of the companies also field-tested their materials. Both companies' instructional materials are standards-based to ensure effectiveness.

# ETHICS MATERIALS IN MODULAR ACTIVITIES

The following section focuses on how well ethics has been incorporated into vendor-generated modules. Similar to the preceding textbook section, vendors were invited to participate in a survey to reveal insight on current ethics instruction within technology education. The purpose of the survey was strictly to identify what is available, not critique vendor-developed materials. Twelve vendors were identified from professional publications and participation in professional conferences, primarily the annual conference of the International Technology Education Association (ITEA). Out of the 12 vendors approached, Applied Educational Systems, Inc., DEPCO, Inc., Hearlihy, and Lab Volt Systems chose to participate in the study.

## Brief Overview of Modular Technology Education

The evolution of technology education out of industrial arts has encompassed a variety of curriculum approaches. According to Warner (1959), technology education "derived via socioeconomic analysis of technology and not by job or trade analysis of the commoner village trades, such as those of the carpenter, the blacksmith, the cabinet maker ..." This approach to teaching the curriculum emphasized the organizing of content on human activity rather than on tool skills. Olson (1963) added that the technology curriculum should include the analysis of the function of the personal life. He interpreted this as comprising anything from one's occupational life to one's recreational life. Accordingly, the curriculum was to emphasize not only human activity, but also tool skills. Thus began the struggle for technology education to define the best approach for teaching its curriculum.

The Industrial Arts Curriculum Project (IACP) embarked on this process in 1968 by initiating the inclusion of "practices used to change materials to add to their worth and the problems associated with creating these changes" (Wright, 1995). Again, the curriculum included both tool skills and human activity. However, for the first time, the curriculum incorporated the impact technology was having on our world. Despite these early curriculum advances, the 1980s proved to be the decade with the most influential changes in the technology education curriculum and its approaches (Dugger and Yung, 1995). Even though the term module had been used to describe individualized learning units since the 1970s (Reed, 2001a), it was not until the mid-1980s when Industrial Arts was changed to

Technology Education that the modular approach to technology education (MATE) began to take root in the technology labs (Dean, 1997).

Early instructional modules were developed as general education tools and were influenced by the Gestalt principle of summation and the teaching machines of B. F. Skinner (Reed, 2001a). Skinner's machines incorporated small instructional steps, active student involvement, immediate confirmation or reinforcement, and self-pacing. Similarly, the MATE requires students to assimilate knowledge from reading, watching videos, and working through software to solve problems utilizing technological tools of today through self-paced, hands-on activities. Accordingly, MATE is defined as ". . . completely (or nearly completely) organized such that students rotate among content modules in which all of the instructional materials and equipment are provided, requiring minimal assistance or instruction from the teacher" (Brusic and LaPorte, 2000, p. 8).

Although only about one-sixth of the technology education labs in the United States were identified as modular in a recent national study, about half of the survey respondents incorporated some form of vendor-generated curriculum (Sanders, 2001). Obviously, modular technology education and vendor curriculum have opened a new venue for technology vendors. For a detailed account of one company's development in this area, Dean (1997) provides a history of PITSCO, Inc., which is one of technology education's most prominent module suppliers.

Hence, another caveat in the ongoing struggle for technology education is to define the best approach for teaching. According to Petrina (1993), ". . . corporate MATEs admit only selected views and ideologies on the social and cultural interaction with technology. Shaped by corporate values and market interests, corporate MATEs basically amount to 'company' views of the technological world; and consequently, determine *what* and *whose* knowledge is legitimate" (italics original, p. 77). This same issue is important when considering whose view of ethics should be taught (Hill and Dewey, 2001). Corporate modules are costly, but they are a contemporary approach for a dynamic program that still covers the basics of technology while producing technologically literate students (R. Barker, personal communication, January 23, 1998).

### *Identifying Included Ethics Topics*

According to the vendor surveys for this chapter, all participants stated that they did produce technology education curriculum materials for grade levels 6–12 that contain instruction on ethics. The types of curriculum materials produced that include ethics were (*a*) textbooks and/or workbooks, (*b*) videos, (*c*) teacher materials, (*d*) student-centered activities (for example, modular materials, kits), and (*e*) software and/or multimedia (for example, CD, DVD, Web-based). Participants indicated a variety of methods in which materials are reviewed to ensure the effectiveness of their curriculum. Each vendor stated that materials were reviewed and field-tested prior to publication and that their materials were standards-based. Two of the respondents indicated that they also received instructor feedback after the materials were implemented, and one vendor also solicited student feedback.

All four companies stated that their materials were based on the *Standards for Technological Literacy: Content for the Study of Technology*. Two of the four stated the materials were also based on the Secretary's Commission on Achieving Necessary Skills (SCANS). Other influencing standards included those published by the International Society for Technology in Education (ISTE), National Science Teachers Association (NSTA), National Council of Teachers of Mathematics, National Coalition for Advanced Manufacturing (NACFAM), AgriScience, in-house (company) standards, and individual state standards.

The same areas from *Standards for Technological Literacy* that were utilized in the textbook survey described previously were also utilized in the module vendor survey. Three of the four companies stated their curriculum moderately (one lesson or activity) covered ethics in the following 15 categories:

1. Characteristics and scope of technology
2. Relationships among technologies and connections between technology and other fields
3. Cultural, social, economic, and political effects of technology
4. The effects of technology on the environment
5. The role of society in the development and use of technology
6. The influence of technology on history
7. Design
8. Problem solving
9. The impacts of products and systems
10. Agricultural and biotechnology

11. Energy and transportation
12. Information and communication
13. Transportation
14. Manufacturing
15. Construction

One of these four companies also stated their materials included ethics instruction as it relates to medical technology. Only one company stated that they comprehensively (several lessons or activities) covered ethics in all 17 categories, including the core concepts of technology (systems, resources, requirements, optimization and trade-offs, processes, and controls).

*Measurement Strategies for Ethics-Related Outcomes*

John Richardson, a forerunner in vendor-generated technology education curriculum, claimed "kids actually experience why they need to learn and they will retain knowledge by seeing and experiencing concrete applications as opposed to memorizing answers to a test" (Potsky, 1997). This statement sums up the assessment strategy for most modular vendors. All four companies surveyed utilized multiple assessment strategies in their MATE. Written work, tests or quizzes, and portfolios were commonly used by all four companies. Concomitantly, only one of the companies stated that they utilized group or class discussion. This is an interesting finding given the social and affective nature of ethics. Other assessment strategies included student presentations, audio or video, electronic activity (for example, Web page, digital video), and projects.

*Survey Influences*

During follow-up of the textbook and modular vendor surveys at the 2003 ITEA conference in Nashville, Tennessee, some interesting feedback emerged. First, several vendors admitted that they did not respond to the survey because they did not feel they were incorporating ethics instruction in their materials. When a discussion ensued regarding copyright, building codes, and other similar concepts covered in technology education, however, many vendors realized they do, in fact, touch on ethics. A second issue emerged when one vendor, upon reviewing the survey, realized they were not incorporating ethics. The survey prompted a review of *Standards for*

*Technological Literacy* and the inclusion of ethics in updates of their materials.

## ETHICS IN TECHNOLOGY EDUCATION LITERATURE

A wide range of sources discuss the importance of ethics in technology education. However, unlike texts and modular materials, standards and scholarly literature often make recommendations but leave instructional methods, activities, and assessments up to curriculum designers.

### Sources

The previous review of texts and modular materials shows the influence of standards as a primary source for curriculum development. The *Standards for Technological Literacy: Content for the Study of Technology* (ITEA, 2000) also plays a key role for teachers, administrators, and teacher educators.

The *Standards for Technological Literacy* addresses ethics in Standard 4 and benchmarks F and J (see Table 6-1). These cognitive benchmarks are unique in their own right, but are closely related. Benchmark F clearly focuses on the identification of ethical issues to help students develop an understanding of the relationship between societal concerns and technology. Benchmark J encourages opportunities for students to build on these cognitive skills by helping them to understand the importance of ethical decisions as citizens who use, manage, and assess technology.

The *National Standards for Social Studies Teachers* (National Council for the Social Studies, 1997) also highlights the importance of decision making and ethics. Specifically, thematic Standard 8 on science, technol-

Table 6-1. Ethics Addressed in the *Standards for Technological Literacy* (ITEA, 2000)

| |
|---|
| **Standard 4:** Students will develop an understanding of the cultural, social, economic, and political effects of technology (p. 57). |
| **Benchmark F (grades 6–8):** The development and use of technology poses ethical issues (p. 61). |
| **Benchmark J (grades 9–12):** Ethical considerations are important in the development, selection, and use of technologies (p. 63). |

ogy, and society emphasizes the need for teachers to help students use ethical standards when analyzing the physical world. Such parallels to the *Standards for Technological Literacy* provide opportunities for teachers to use interdisciplinary, collaborative experiences with their students (Sanders and Binderup, 2000).

Equally important to standards-based curriculum is the scholarly research that supports curriculum and teaching practice. Unfortunately, research on ethics in technology education is limited. A search of over 5,200 technology education theses and dissertations spanning more than 100 years only turned up two graduate studies related to ethics and these focused on the topic of work ethic (Reed, 2001b). Nevertheless, there are notable studies and instructional materials in the professional literature.

## *Topics Covered*

A great deal of the technology education literature regarding ethics stresses the need for teachers to include the social context inherent in science and technology studies (STS). For example, many technology educators have traditionally focused on the impacts of technology as if there are no social influences on these forces. A more holistic instructional approach, however, includes the concept that creators, users, and consumers of technology shape the direction of those impacts as well as the technologies themselves (Pannabecker, 1991).

To teach these STS concepts, less time is spent on technical skills and concepts and more time on social content. Computer-aided design (CAD), traditionally a skill-driven subject, can provide an excellent example of how to balance technical and social content. A sociotechnical CAD class would include many ethical-personal topics, including, among others, the sociology, psychology, history, and economics associated with CAD (Petrina, 2003).

A sociotechnical approach is also helpful for teaching ethics related to biotechnology, which is a relatively new topic in technology education. By looking at the environmental and cultural issues associated with these technologies, in addition to the economic and technical aspects, students would be better prepared to make decisions that are global in scope (Conway, 2000). Topics with strong support for instruction on bioethics include social impacts, principles of ethics, impacts of using biotechnology, regulation (legislation and safety), and potentials of gene therapy (Wells, 1994).

Identifying instructional content based upon research and published sources is important. Some areas, such as appropriate technology, provide an abundance of material on ethics. Appropriate technology topics suitable for instruction include, among others, environmental pollution, labor issues, and nonrenewable energy sources (Hill and Dewey, 2001). Other technology topics with ethical considerations are public response to information technology, access to medical care for the aging population, and technological control of the environment (Hendricks, 1996).

## *Context and Instructional Activities*

The National Academy of Engineering and National Research Council (2002) claims that a technologically literate person should develop ways of thinking and acting along with acquisition of technological knowledge and skills. A significant part of these "ways of thinking" involves the ability to ask questions, seek information, and make decisions. Each of these skills has implications for instruction on ethics. But how can technology educators effectively implement these sociotechnical skills outlined by content standards, STS, and professional organizations?

Obviously, the ideal method is to have a technology education course dedicated to ethics. Although this might not be practical in a secondary school setting, Todd and Karsnitz (1999) demonstrated how this could be accomplished at the university level. The *Society, Ethics, and Technology* course developed at the College of New Jersey was implemented to strengthen the general education requirements. The course was approved based on the rationale that technology had become such an important part of society that a true liberal arts education now required knowledge of this complex relationship.

A second source of strategies for teaching sociotechnical skills can be found by examining other disciplines. Preparation for medical professions, for example, includes instruction in three areas: (*a*) skills/practice, (*b*) knowledge/information, and (*c*) values/ethics. Figure 6-1 illustrates the balance of this triangular association utilized in medicine. Like medical instruction, technological studies could start at any one of these points and progress to the other two (Gradwell, 1999). This trinity is representative of the characteristics of a technologically literate person as outlined by the National Academy of Engineering and National Research Council (2002) and includes knowledge, ways of thinking and acting, and capabilities.

Figure 6-1. Ensuring balance in technology (Gradwell, 1999).

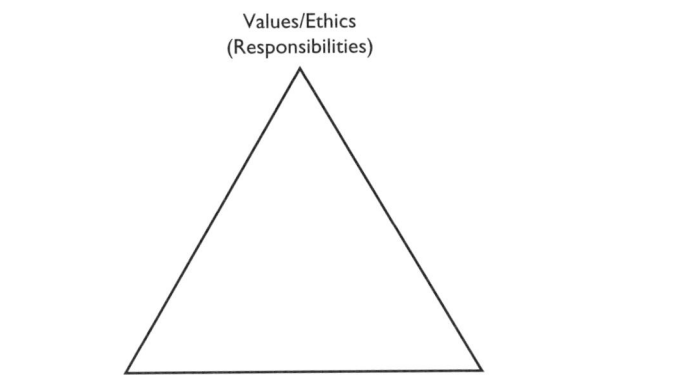

Interdisciplinary activities can incorporate philosophical activities. For example, one interesting exercise is to have students view technology as a prosthesis or extension of the human body. Comparisons relating technology directly to the human body can help issues become more personalized, thus causing students to ask themselves tough questions and become more engaged in their thinking (Huyke, 2001).

The ITEA has provided an additional source for guidance in implementing ethics material and sociotechnical instruction. Monographs on interpreting the *Standards for Technological Literacy* (Meyer, 2000a) and assessment strategies (Meyer, 2000b) have been developed to help teachers create curriculum that is tied to research. The ITEA's Center for the Advancement of Teaching Technology and Science (CATTS) also has created curriculum materials and training to help teachers implement standards-based instruction.

Curriculum materials developed by teachers and teacher educators also have tremendous potential for getting ethics instruction into the technology education classroom. Wells' (2000) *Technology Education Biotechnology Curriculum* is an excellent example that includes a framework for objectives and learning outcomes as well as activities on bioethics. In a second example, Gorman (1998) used case studies, simulation, and learning modules to teach secondary students about the integral role of ethics in discovery and invention.

# INTERNATIONAL TECHNOLOGY EDUCATION AND ETHICS

The previous professional literature discussion reflected methods for incorporating ethics instruction into technology education. Many countries, however, already have varying degrees of instruction on ethics. Although a complete survey of international instruction on ethics in technology education is beyond the scope of this chapter, this section highlights some of the more significant examples.

## Extent of Ethics Instruction

International instruction on ethics generally comes in three forms: (*a*) courses for students, (*b*) materials for teacher education, and (*c*) teacher resources. Although some countries require students to take ethics courses, these are not necessarily specific to technology education. Korea, for example, requires academic high school students to take a course in ethics (Yi, 1997).

Teacher education materials on ethics generally fall into two categories: (*a*) a reflection of STS practice or (*b*) the standards and curriculum that future teachers will implement after they enter the classroom. Incorporating STS methods requires the teacher educator to incorporate ethics materials and highlight the importance of sociotechnical instruction into their courses. Often, the process is an evolutionary one that requires the teacher educator to continually evaluate the curriculum as well as sociotechnical trends (Petrina, 2003). Incorporating ethics instruction based on local standards and curriculum is a common practice. However, whether the preservice instruction on ethics is carried over to primary, elementary, and secondary classrooms is questionable (K. Volk, personal communication, July 31, 2002).

Resources for the classroom teacher are the primary source of international instruction on ethics. Many countries provide teachers with standards, objectives, and curriculum materials. However, the depth of these materials as they relate specifically to ethics varies widely. The best way to understand the range of resources is to review implementation strategies.

## Strategies for Including Ethics in Technology Education

Some countries are working on initiatives to bring technology education courses into the mainstream curriculum. Finland, for example, has a core curriculum that allows local flexibility for subjects like technology education. Recent developments, however, are creating national interest for consistency in technology education classes (Järvinen and Nykänen, 2002). For now, ethics instruction is left up to the technology teacher. Finnish technology teachers are required to take courses in educational science and ethics as a part of their preservice studies (Alamäki, 2000), so influences are in place to encourage inclusion of ethical issues in instruction.

Countries with more established technology education programs, such as New Zealand, England, Canada, and Australia, have published materials for including ethics in technology education. New Zealand created a document titled *Technology in the New Zealand Curriculum* (Ministry of Education) in 1995 that contains three strands: (*a*) technological knowledge and understanding, (*b*) technological capability, and (*c*) technology and society (Reid, 2000). The third strand contains an achievement objective stating that students "should develop awareness and understanding of the beliefs, values, and ethics of individuals and groups" (Ministry of Education, 10). This comprehensive document explains each achievement objective and provides learning and assessment examples for each level in secondary education.

England has developed a scheme of work that is the overall plan of design and technology. The scheme of work is made up of units that are usually designed to be completed over a term or less. Units set out specific learning objectives that reflect the program of study, as well as possible teaching activities and learning outcomes (Department for Education and Skills 2002). A search of these units indicated that inclusion of ethics is primarily limited to the secondary level, years seven through nine.

Several Canadian provinces also have materials for teachers that incorporate ethics instruction. In British Columbia, one of the primary goals of technology education is to help students "develop the ability to deal ethically with technology" (British Columbia Education, 2002). An Integrated Resource Package is available to help teachers integrate STS concepts, including ethics. Saskatchewan also provides teachers with a handbook that includes models detailing how to discuss social issues (Saskatchewan Education 2002).

In Australia, Williams (2001) conducted a comprehensive study of technology education curriculum as it related to values. Documents from all eight states and territories were reviewed as well as data from interviews and focus groups. The term *values* was broken down into five areas, including social, cultural, environmental, economic, and ethical. One of the key findings was that all five areas were generally not being addressed effectively (Williams).

*Outcomes Identified and Achieved*

Many of the international materials highlighted in the preceding section are curriculum guides or units that do not contain specific outcomes for ethics. Materials usually include ethics as one of the objectives or part of a unit. Seven of the eight Australian states and territories, however, did have outcomes for ethics, although Williams (2001) revealed that the inclusion of those particular outcomes was limited and not systematic. For example, several states and territories included ethics in the primary years but the majority of material was found in the senior years. The most comprehensive attention to ethics was found in instructional technology and primarily dealt with privacy, hacking, intellectual property, and security. Broader areas in which ethical issues were addressed included gene technology, agriculture, designing, copyright, design for the disabled, and food technology (Williams).

*Availability of Materials for Adoption or Use by Others*

Some of the international materials discussed in this chapter are available on the World Wide Web. Uniform Resource Locators (URLs) are listed in the reference section of this chapter for the Canadian provinces of British Columbia (British Columbia Education, 2002) and Saskatchewan (Saskatchewan Education, 2002). England also has a Web site for the design and technology curriculum that contains a teacher's guide and instructional units for the primary and secondary levels (Department for Education and Skills, 2002). New Zealand's document titled *Technology in the New Zealand Curriculum* is available in print from the Ministry of Education (1995).

## PROFESSIONAL ETHICS IN TECHNOLOGY EDUCATION

The ethical influence of technology educators at all levels should emanate from their professional knowledge and behavior rather than from the power derived from their role. This section discusses the importance for technology education teachers, administrators, and teacher educators to demonstrate ethical behavior at all times. Technology education literature on the significance of modeling ethical behavior and aspects for teacher education is limited. However, professional development standards and materials from other disciplines and organizations have laid the foundation for expanding ethics-related resources in technology education.

### Ethical Behavior by Technology Education Teachers

The role of technology teachers in encouraging high ethical standards is of utmost importance because they interact directly with students from a wide range of cultural, social, and economic backgrounds. In an age in which the political climate is holding teachers more and more accountable through accreditation, standardized testing, and certification, it is imperative for teachers to display ethical behavior in and out of the classroom. Ethical education should be an integral part of teacher preparation and in-service education.

In a survey by Wiens (1996), technology teacher education coordinators realized the importance of incorporating sociocultural aspects of technology, including ethics. However, many of the respondents realized these aspects were not being addressed despite their inclusion in the National Council for Accreditation of Teacher Education (NCATE) guidelines. Meeting NCATE guidelines by including sociocultural instruction in preservice education serves two important functions. First, such instruction can provide students with methods and materials for incorporating ethics into their future technology education programs. Humanities courses in ethics can also benefit preservice teachers by teaching them how to address technological decisions (Wiens, 1995). Second, preservice teachers can learn the importance of modeling ethical behavior in the classroom. The behavior used in the classroom should be exemplary and consistent with professional standards.

The Technology for All Americans project created assessment standards, professional development standards, and program standards to help teachers after they entered the classroom. The narrative materials for professional development standard PD-6 states "Pre-service and in-service professional development experiences should prepare teachers to engage in comprehensive and sustained personal professional growth" (ITEA, 2003, p. 60). Ethical behavior is exclusively mentioned under PD-6, guideline B: Teacher candidates and educators should "establish a personal commitment to ethical behavior within the educational environment as well as in private life" (p. 60).

## Ethical Behavior by Technology Education Administrators

Administrators generally do not have as much contact with students as teachers, but should adhere to the same ethical standards. Also, because administrators deal more directly with other administrators, parents, school boards, and community leaders, they should be cognizant of professional standards. The National Education Association (NEA) highlighted this need in 1975 when it created a code of ethics for education that included two provisions: "Commitment to the Student" and "Commitment to the Profession" (Wagner, 1996).

A more direct example of professional commitment for technology educators is provided by ITEA. The ITEA has adopted standards of conduct for the technology education profession. All members, including students, teachers, administrators, and teacher educators, are expected to adhere to the code of conduct, which includes actively promoting and encouraging "the highest level of ethics within the technology teaching profession" (Devier 1999, 255). The ITEA has also implemented a recognition program for professionals who have demonstrated significant leadership. The guidelines of the Distinguished Technology Educator (DTE) award include several leadership attributes, such as modeling "behaviors and characteristics that include ethical, moral, and legal aspects of being a professional educator" (Wright 1999, 186).

## Ethical Behavior by Technology Teacher Educators

Teacher educators have several significant reasons to incorporate instruction related to ethics into their programs. The obvious reason is to portray ethical behavior and provide preservice teachers with instructional methods and materials. A second, more long-range goal is to provide students with the ability to connect classroom experiences with higher-order outcomes. Developing an independent learning ethic helps teachers to see the value of lifelong learning (Hanson 1993). The importance of these concepts has been recognized by the profession through provisions in the CTTE/ITEA/NCATE guidelines for technology teacher preparation programs.

The CTTE/ITEA/NCATE guidelines for technology teacher education stipulate that each of the 10 standards for teacher preparation programs should be explained by outcome statements called indicators. Knowledge, performance, and disposition indicators are included with each standard. Disposition indicators are those that deal with attitudes, values, ethics, beliefs, and affective behaviors in relation to the standard.

It is important to remember that when writing a folio, it is not necessary to respond to each and every indicator. The folio and supporting documentation should be written to the standards. However, mastery of indicators will lead to more complete achievement of each standard. (Council on Technology Teacher Education [CTTE], 2002, Section III-12)

These guidelines highlight the importance for teacher education programs to document how they are including ethics in their preservice courses.

## *Aspects of Ethics to be Included in Teacher Education*

The majority of materials reviewed to this point have outlined the need for instruction in ethics. But whose view of ethics should we be using in technology teacher education? Sensitivity in the selection of instructional materials is extremely important. Hill and Dewey (2001) presented three possible methods for including ethics: (*a*) selecting topics that represent commonly agreed upon concepts, (*b*) utilizing published work, or (*c*) taking a research-based approach. All three methods should be introduced in teacher education even though the last two are more defensible (Hill and Dewey).

The first approach to ethics instruction, utilizing commonly agreed upon concepts, is a good approach to foster classroom discussion. The corporate scandals in the United States at the beginning of the twenty-first century provide a good example. A wide range of media commentaries from that time could be used to discuss the importance of ethical behavior by professionals. This approach is good for creating awareness on ethics, but care should be taken to help students become aware that actions adopted within an organization are not necessarily ethical. The basis for ethics must extend beyond approval by a particular subset of society.

Organizations such as the Josephson Institute of Ethics have developed materials on ethics that can be included in teacher education. Strategies outlined by the Josephson Institute help students to identify ethical issues as well as to acquire skills on how to take action. The first aspect involves the ability to discern right from wrong, good from evil, and propriety from impropriety. The second involves the commitment to do what is right, good, and proper (Josephson Institute of Ethics, 2002a). To help implement these components into education, the Josephson Institute runs a project known as *Character Counts!* that has several hundred coalition members. Members include a variety of organizations, school systems, communities, cities, and counties.

The *Character Counts!*, coalition is designed "to fortify the lives of America's young people with consensus ethical values called the 'Six Pillars of Character'" (Character Counts!, 2002). The six values include trustworthiness, respect, responsibility, fairness, caring, and citizenship. Space here does not permit a complete discussion on how to incorporate these six values into technology teacher preparation. However, *Character Counts!* does provide information on its Web site regarding materials, awards, and training (http://www.charactercounts.org/).

Research can also provide materials for inclusion in teacher preparation. For example, a survey conducted by the Josephson Institute in 1998 is helpful to show preservice teachers what secondary students in the United States believe about ethical behavior. Compelling data is also presented on whether these students emulated ethical behavior with family, friends, and the public (Josephson Institute of Ethics, 2002b). In a narrower survey dealing strictly with work ethic, Hill (1995) identified three constructs that are useful for teacher educators: interpersonal skills, initiative, and being dependable. These examples show how teacher educators can use research to emulate ethical behavior, provide instructional methods and materials, and foster a learning ethic that promotes lifelong learning.

Additional examples of incorporating ethics were presented in Chapter 1. The list of attributes provided (integrity, responsibility, fairness, caring, initiative, interpersonal skills, and dependability) are based on research and are universally accepted. The model outlined in Chapter 1 (see Figure 1-1) also ties to the problem-solving model that has traditionally been used in technology education courses.

## SUMMARY

How far have we come? This initial question was posed to see how well the technology education profession has incorporated ethics to help advance technological literacy. Five areas were reviewed: (*a*) vendor textbooks and print materials, (*b*) vendor modules, (*c*) professional literature, (*d*) international ethics materials, and (*e*) professional ethics for technology teachers, teacher educators, and administrators.

In the first two sections, a survey was completed with textbook companies and module suppliers to identify commercial materials that incorporated instruction on ethics. Respondents indicated that ethics was incorporated into a broad range of technology topics using a variety of delivery methods and a mixture of assessment strategies. The *Standards for Technological Literacy* (ITEA, 2000), as well as other standards, has an impact on the development of materials.

The professional literature reviewed on ethics in technology education indicated strong support for its inclusion; however, there was very little substantive research. Much of the literature was in the form of editorials, standards, and curriculum frameworks. A similar trend was found in the international community. Ethics was often one objective or part of a unit or lesson rather than the central focus. Clearly, there is room for more research on the inclusion of ethics in technology education and a need for curriculum development. Materials are available from organizations that focus on ethics; however, these are not specific to technology education. These materials do provide a significant opportunity for interdisciplinary instruction, especially because the importance for instruction on ethics as it relates to technology is highlighted in standards from multiple disciplines.

Professional ethics for technology teachers, teacher educators, and administrators is a third area that needs further development. Much of the literature in technology education has stressed the need for professional educators to emulate ethical behavior at all times, but guidelines or examples are usually not provided. This situation demonstrates the paradox of our initial question, *how far have we come?* Clearly, we have established the baseline for ethics within technology education. However, the horizon changes with each step. Out of all the disciplines in modern education, technology education has traditionally been the leader in experiential learning. We now have an opportunity to also become the leader for sociotechnical instruction by incorporating ethics into our classrooms, laboratories, and through interdisciplinary collaborations.

## REFLECTION QUESTIONS

1. To what extent do companies that develop textbooks, modules, and other instructional materials shape the nature of technology education content?
2. This chapter has provided an overview of existing materials that support ethics instruction as a component of technology education. In

a situation in which modular materials are being used to teach technology education, how should ethics instruction be handled if not included in the modular materials?
3. How can technology educators influence companies to include a greater emphasis on ethics in the instructional materials they produce?
4. To what extent was ethics included in the content of industrial arts?
5. How important is it for teacher preparation programs to include a course related to society, ethics, and technology within their programs of study?
6. What can technology educators in the United States learn from their international counterparts about including ethics within the study of technology?
7. How important is professional ethics within the field of technology education? If teachers failed to abide by copyright laws, displayed a poor work ethic, or demonstrated poor judgment in their decisions, what would the impact be on their students?
8. How would you approach preparation of a technology education activity that encourages the development of good character traits?

# REFERENCES

Alamäki, A. (2000). Current trends in technology education in Finland. *Journal of Technology Studies, 26*(1), 19-23.

British Columbia Education. (2002). *Technology education kindergarten to grade 12 objectives.* Retrieved December 5, 2002, from http://www.bced.gov.bc.ca/irp/te11_12/intro3.htm

Brusic, S. A., & LaPorte, J. E. (2000). The status of modular technology education in Virginia. *Journal of Industrial Teacher Education, 38*(1), 5-28.

Character Counts! (2002). *Character Counts!* Retrieved December 5, 2002, from http://www.charactercounts.org/

Conway, R. (2000). Ethical judgments in genetic engineering: The implications for technology education. *International Journal of Technology and Design Education, 10*(3), 239-254.

Council on Technology Teacher Education (CTTE). (2002). *Draft ITEA/CTTE/NCATE standards.* Reston, VA: Author.

Dean, H. R. (1997). *Changing education: A success story.* Dallas, TX: Arete.

Department for Education and Skills. (2002). *The Standards Site: Schemes of Work.* Retrieved December 5, 2002, from http://www.standards.dfes.gov.uk/schemes/

Devier, D. H. (1999). Fostering a professional culture in technology education. In A. F. Gilberti & D. L. Rouch (Eds.), *Advancing professionalism in technology education* (pp. 251-270). Peoria, IL: Glencoe/McGraw-Hill.

DeVore, P. W. (1980). *Technology: An introduction.* Worcester, Massachusetts: Davis Publications, Inc.

DeVore, P. W. (1991). Technological literacy: The evolving paradigm. In M. J. Dyrenfurth & M. R. Kozak (Eds.), *Technological Literacy* (pp. 251-279). Peoria, IL: Glencoe/McGraw-Hill.

Dugger, W. E., & Yung, J. E. (1995). *Fastback: Technology education today* (308th ed.). Bloomington, IN: Phi Delta Kappa.

Gorman, M. E. (1998). *Transforming nature: Ethics, invention, and discovery.* Boston: Kluwer Academic.

Gradwell, J. B. (1999). The immensity of technology ... and the role of the individual. *International Journal of Technology and Design Education, 9*(3), 241-267.

Hanson, R. E. (1993). A technological teacher education program planning model. *Journal of Technology Education, 5*(1), 21-28.

Hendricks, J. L. (1996). Technology—Social and interpersonal interaction. In R. L. Custer & A. E. Wiens (Eds.), *Technology and the quality of life* (pp. 321-344). Peoria, IL: Glencoe/McGraw-Hill.

Hill, R. B. (1995). A new look at selected employability skills: A factor analysis of the Occupational Work Ethic. *Journal of Vocational Education Research, 20*(4), 59-73.

Hill, R. B., & Dewey, G. (2001). Moral and ethical issues related to appropriate technology. In R. C. Wicklein (Ed.), *Appropriate technology for sustainable living* (pp. 78-91). Peoria, IL: Glencoe/McGraw-Hill.

Huyke, G. J. (2001). Toward an ethics of technologies as prostheses. *International Journal of Technology and Design Education, 11*(1), 53-65.

International Technology Education Association (ITEA). (2000). *Standards for technological literacy: Content for the study of technology.* Reston, VA: Author.

International Technology Education Association (ITEA). (2003). *Advancing excellence in technological literacy: Student assessment, professional development, and program standards.* Reston, VA: Author.

Järvinen, E-M., & Nykänen, J. (2002). *Technology education now!* Paper presented at the International Technology Education Association annual conference, Columbus, OH.

Josephson Institute of Ethics. (2002a). *Making ethical decisions—What is ethics anyway?* Retrieved December 5, 2002, from http://www.josephsoninstitute.org/MED/MED-whatisethics.htm

Josephson Institute of Ethics. (2002b). *1998 report card on the ethics of American youth.* Retrieved December 5, 2002, from http://www.josephsoninstitute.org/98-Survey/98survey.htm

Meyer, S. (2000a). *Criteria for Interpreting the Standards for Technological Literacy: Content for the Study of Technology.* Reston, VA: International Technology Education Association.

Meyer, S. (2000b). *Assessment Strategies for the Standards for Technological Literacy: Content for the Study of Technology.* Reston, VA: International Technology Education Association.

Ministry of Education. (1995). *Technology in the New Zealand Curriculum.* Wellington, New Zealand: Learning Media Limited.

National Academy of Engineering and National Research Council. (2002). *Technically speaking: Why all Americans need to know more about technology.* In G. Pearson & A. T. Young (Eds.), *Committee on Technological Literacy.* Washington, DC: National Academy Press.

National Council for the Social Studies. (1997). *National Standards for Social Studies Teachers.* Washington, DC: Author.

Olson, D. W. (1963). *Industrial arts and technology.* Englewood Cliffs, NJ: Prentice-Hall.

Pannabecker, J. R. (1991). Technological impacts and determinism in technology education: Alternate metaphors from social constructivism. *Journal of Technology Education, 3*(1).

Petrina, S. (1993). Under the corporate thumb: Troubles with our MATE (Modular Approach to Technology Education). *Journal of Technology Education, 5*(1), 72-80.

Petrina, S. (2003). Two cultures of technical courses and discourses: The case of computer aided design. *International Journal of Technology and Design Education, 13*(1), 47-73.

Potsky, A. (1997, February). Buffet style learning. *Techniques,* 40-43.

Reed, P. A. (2001a). Learning style and laboratory preference: A study of middle school technology education teachers in Virginia. *Journal of Technology Education, 13*(1), 59-71.

Reed, P. A. (Ed.). (2001b). *The technology education graduate research database: 1892–2000.* Retrieved December 5, 2002, from http://www.teched.vt.edu/CTTE/

Reid, M. S. (2000). Towards effective technology education in New Zealand. *Journal of Technology Education, 11*(2), 33-47.

Sanders, M. E. (2001). New paradigm or old wine? The status of technology education in the United States. *Journal of Technology Education, 12*(2), 35-55.

Sanders, M. E., & Binderup, K. (2000). *Integrating technology education across the curriculum.* Reston, VA: International Technology Education Association.

Saskatchewan Education. (2002). *Understanding the common essential learnings: A handbook for teachers.* Retrieved December 5, 2002, from http://www.sasked.bov.sk.ca/docs/policy/cels/

Todd, R. D., & Karsnitz, J. R. (1999). Identifying and solving professional problems. In A. F. Gilberti & D. L. Rouch (Eds.), *Advancing professionalism in technology education* (pp. 139-164). Peoria, IL: Glencoe/McGraw-Hill.

Wagner, P. A. (1996). *Fastback: Understanding professional ethics* (403rd ed.). Bloomington, IN: Phi Delta Kappa.

Warner, W. E. (1959). *The industrial arts curriculum: Development of a program to reflect American technology.* Columbus, OH: Epsilon Pi Tau.

Wells, J. G. (1994). Establishing a taxonometric structure for the study of biotechnology in secondary school technology education. *Journal of Technology Education, 6*(1), 58-75.

Wells, J. G. (2000). *Technology education biotechnology curriculum.* Morgantown, WV: TEBC Project.

Wiens, A. E. (1995). Technology and liberal education. In G. E. Martin (Ed.), *Foundations of Technology Education* (pp. 119-152). Peoria, IL: Glencoe/McGraw-Hill.

Wiens, A. E. (1996). Teaching about socio-cultural dimensions of technology development and use. *The Technology Teacher, 55*(6), 23-26.

Williams, P. J. (2001). *The teaching and learning of technology in Australian primary and secondary schools.* Department of Education, Science and Technology Working Report, Commonwealth of Australia, 299 pp.

Wright, R. T. (1995). Technology education curriculum development efforts. In G. E. Martin (Ed.), *Foundations of Technology Education* (pp. 247-285). Peoria, IL: Glencoe/McGraw-Hill.

Wright, J. R. (1999). Teacher professionalism in higher education. In A. F. Gilberti & D. L. Rouch (Eds.), *Advancing professionalism in technology education* (pp. 181-198). Peoria, IL: Glencoe/McGraw-Hill.

Yi, S. (1997). Technology education in Korea: Curriculum and challenges. *Journal of Technology Studies, 23*(2), 42-49.

# Ethics and the Study of the Designed World

## Chapter 7

Michael A. DeMiranda
Colorado State University
Fort Collins, CO

Richard D. Seymour
Ball State University
Muncie, IN

Nick Benson
Colorado State University
Fort Collins, CO

Jack Wescott
Ball State University
Muncie, IN

Len S. Litowitz
Millersville University
Millersville, PA

Myra N. Womble
The University of Georgia
Athens, GA

Mark Sanders
Virginia Tech
Blacksburg, VA

Stephanie Williams
The University of Georgia
Athens, GA

The field of technology education has always been known for use of hands-on experiences in teaching. In recent years, the importance of context has been increasingly recognized as an important facet of successful instruction. Successfully addressing ethical issues requires that topics be addressed within real-world scenarios and associated with related technical content. This approach would likely result in greater success than would result from isolated lessons on ethics.

In this chapter, there are several examples of instructional activities that could be integrated into medical technologies, agricultural and related biotechnologies, energy and power technologies, information and communication technologies, transportation technologies, manufacturing technologies, and construction technologies instruction. Design and problem solving are integrated into the activities described and some of the problems presented include embedded ethical dilemmas. This section addresses multiple items from the *Standards for Technological Literacy* (International Technology Education Association, 2000), including Standard 11, apply the design process, and Standard 13, assess the impact of products and systems.

## MEDICAL TECHNOLOGIES
*Michael A. DeMiranda and Nick Benson*

Early doctors were not able to hear a patient's heartbeat or lungs very clearly. They simply put an ear to a person's chest and listened. This was not always possible or ethical when doctors were examining female patients. Faced with this dilemma, one doctor, when presented with a female patient, tried something new. He remembered that sound travels in all directions but can become trapped and reflected in a tube. Using problem-solving skills, he developed a solution.

Rene Laennec created the first stethoscope by rolling up a pile of paper. He then modified the design using a wooden tube with a funnel-shaped opening at one end to place against the body. The stethoscope went through many different modifications to become what you see in your doctor's office today. Different designs allowed doctors and medical professionals to hear different organs and bodily functions.

Understanding how sound works was an important part of the development of the stethoscope. Sound is created when an object vibrates. Consider plucking a guitar string. When the string is plucked, it moves back and forth. When the string is vibrating toward you, it is pushing the air molecules around it, and they, in turn, are pushing the ones in front of them, and so on, so that they are compacted together approaching your ear. This leads to an increase in pressure, called compression. When the string is vibrating away from you, it is pulling some of the air molecules with it. This results in a decrease in pressure, called rarefaction. The fluctuations between compression and rarefaction are perceived as sound. The rate at which the fluctuations occur, called frequency, affects whether the pitch we hear sounds high or low, whereas how strong the wave is, the amplitude, determines how loud the sound is. See http://www.medimagery.com/ear/how.html for a nice illustration of how sound waves enter the ear canal and vibrate against the tympanic membrane.

Sound waves can actually travel along a string that is attached to a cup on each end, when the string is held taut and straight enough. The vibrations created by speaking into one of the cups travel along the string to the other cup. Hence, the bottom of the second cup begins to vibrate back and forth just like the bottom of the first cup, producing sound waves. This is a simple experiment, but it illustrates an important principle. Sound waves need some sort of medium or matter to travel through, such as air or water.

Sound can also travel through the earth. During some earthquakes, the waves from the quake travel through the earth, hit the surface, and then cause the air to move as well. This is usually heard as a low rumble or boom. In space or some other place in which there is a vacuum, there is nothing for the waves to travel through, so there is no sound. The loud explosions from combat in space during science-fiction films are actually not very realistic!

Sound travels as waves in all directions from its source in concentric spheres, with the intensity dropping off as the area of the sphere it covers grows. If the waves are confined to an area such as a pipe or tube, however, they are trapped and reflected back and forth within the pipe until reaching the other side. Because they do not spread out, but stay in the same area, the intensity does not drop off as it does when waves are not confined. Prisoners on Alcatraz used to communicate with each other through the water pipes, an idea that works using the same concept. Variations of this idea can be used in technology education activities to demonstrate how stethoscopes and other similar devices operate.

## SOUND IN A BOX—GRADE LEVELS 3–7

### *Description*

Students design a listening device that will allow them to listen to the sound generated by a specific object inside of a box.

### *Educational Outcomes*

1. Students will explore the movement and properties of sound waves as they design a listening device.
2. Students will explore ways to trap and reflect sound waves so that they can be heard clearly from a distance.
3. Students will gain a firsthand experience of the design processes used by scientists, technologists, and engineers.

### *Materials List*

- Plastic soda bottles (precut the top parts so kids can use them as funnels)

- Paper towel rolls
- Stiff paper
- Straws
- Plastic tubing (varying sizes)
- Masking tape
- Aluminum foil
- Plastic wrap
- Rubber bands
- Rubber balloons
- Clay
- Children's scissors
- Plastic or paper drinking cups
- String
- Metronome (ticking clock or kitchen timer works well)
- Shoe box (or other box that will house metronome effectively)
- Padding for box (for example, bubble wrap, foam, Styrofoam peanuts)

<u>*Design Challenge*</u>

A group of medical technologists have been assigned the task of designing a listening device that will be used to hear a patient's heart while exercising. Because the physicians who will use your design will not want to get hit by a stray leg or arm, you need to make sure that your design can be held at least a foot away from the patient while listening to his heart. You have been asked to develop a prototype device that will enable a listener to hear an object inside a box.

The portion of the prototype device being held to a listener's ear must be held at least a foot away from the testing box. Only the materials provided can be used in constructing the prototype device. The effectiveness of the prototype design will be determined by testing to see whether a sound being produced inside a box can be identified and accurately described by a listener.

## Testing

Turn the object (for example, metronome) on and place it in the box, surrounded by padding. If the area is very quiet, use a fan or radio to mask the sound coming from the box so that the object cannot be heard by simply listening. Students should use their device to listen to the object inside the box while keeping their ears at least one foot from the box.

## Reflection Questions

1. What is the best shape and size for the part of a listening device to be pressed against the chest (box)?
2. Is it better to cover the object against the chest (box) with a membrane?
3. How would your design work if you held it farther away from the box? (You wouldn't hear the sound as well.) Why? (Some of the sound waves would escape and move away from the device.)
4. How would different types of materials work as the membrane and why?
5. What type of tube is best for connecting to your ear (big, small, thick, or thin)?

# THE STETHOSCOPE—GRADE LEVELS 8–9

## Description

Students design a stethoscope that will allow them to listen to the sounds of the human body.

## Educational Outcomes

1. Students will explore, research, design, and test an early medical technology called a stethoscope.
2. Students will explore ways to trap and reflect sound waves so that they can be heard clearly from a distance.
3. Students will gain experience with the design processes used by scientists, technologists, and engineers.

## Ethics and the Study of the Designed World

### Materials List

- Sixty centimeters of clear, vinyl tubing
- One standard 35mm film container
- One 'y' connector (available at auto part stores for vacuum hose connections)
- One standard party balloon (membrane to cover end of film container)
- One rubber band (used to hold balloon in place on film container)

### Design Challenge

Design and make a working model of a stethoscope using the supplied materials. Test your finished stethoscope to see that it works.

### Testing

Demonstrate a procedure for testing the completed stethoscope—place the film container end against the chest or abdomen and have another person listen through the ends of the tubing. Activities such as measuring heart rate can be performed using the completed instrument.

### Reflection Questions

1. What are the advantages of being able to diagnose illnesses by listening to the internal sounds of the body?
2. Explain how the stethoscope had an impact on medical practice after it was invented by Laennec in 1816.
3. How has the design of stethoscopes changed since the instrument was first invented?
4. Describe the principles of acoustics that cause the stethoscope to work.

## BIOTECHNOLOGIES

*Michael A. DeMiranda*

The world's population is growing continually and, with it, the demand for food. The use of traditional plant protection agents, fertilizers, and breeding will only be able to provide limited help. Biotechnology methods have the power to lower the costs of food production, to increase yield, and to produce high-value, biotechnological products. The future of biotechnology will involve solutions to the problems faced by all stakeholders in the agriculture value chain, including farmers, grain and food processors, food manufacturers, and consumers.

Figure 7-1 illustrates the relationship of science, engineering, and technology in the field of biotechnology. In the case of biotechnologies used in agriculture, specializations are further focused on that area of application. The Venn diagram, however, could be applied to biotechnologies in other areas as well.

## A CHANCE FOR A FUTURE— GRADE LEVELS 11–12

*Description*

Students prepare and present materials that portray the positive aspects of a biotechnology company that might be used to educate citizens in a community about this sector of the designed world.

*Educational Outcomes*

1. Students will be able to identify, discuss, and provide examples of biotechnology applications from the four major categories of biotechnologies.
2. Students will be able to compare agricultural technologies to organic methods.
3. Students will be able to explain the role of pharmaceutical biotechnologies in health promotion.
4. Students will be able to identify and analyze applications of industrial biotechnology.

Figure 7-1. Relationship of science, engineering, and technology in biotechnology.

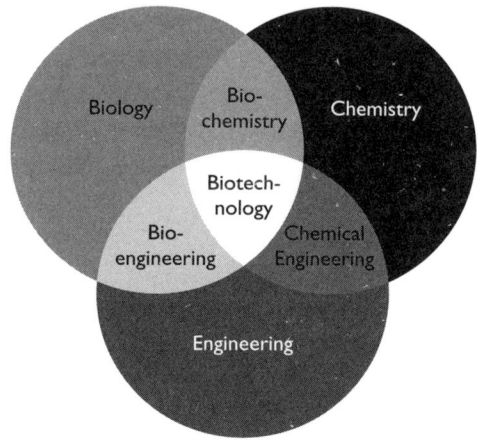

5. Students will be able to provide three examples of biogenetics or bio-engineered products.
6. Students will be able to design a simulation using environmental applications of biotechnology.
7. Students will be able to design, develop, and present a report on a biotechnology of his or her choice using multimedia technology and post the information on the class Web site.
8. Students will use and apply concepts of biotechnology through activities, Web resources, and hands-on laboratory activities.

<u>Materials List</u>

- Cameras
- Computers
- Digital camera
- Internet connection
- Printer
- Projection system
- Laboratory equipment
- Desktop publishing software

- E-mail
- Internet Web browser
- Multimedia software
- Web page development software
- Word processing software

## Design Challenge

You are an engineer at BASF® Corporation in charge of research and development. Your company is planning to relocate to a city 125 miles away from your current location. The purpose of the expansion and relocation is to consolidate the company's industrial biotechnology division. The city council of the new city where BASF® is moving has asked you to present a community briefing on what industrial biotechnologies are. As the lead biotechnologist, your supervisor has asked you to make the presentation.

Your task is to assemble a team of three researchers, and to design, prepare, and publish a multimedia presentation, a four-page color information sheet, and a Web site that will aid the citizens of the community to understand what biotechnologies are and specific applications of industrial biotechnologies. Your persuasive and informative communication must present the ethical, social, and societal benefits of your company's activities.

Approximately six weeks will be provided for completion of this learning activity. Resources will be available to assist you and your team both in the technology education classroom as well as in the media center.

## Testing

Each team will be provided an opportunity to make a presentation to the class as well as to visit members of the advisory council. Members of the audience will critique presentations and provide constructive criticism.

# ENERGY AND POWER TECHNOLOGIES

*Len S. Litowitz*

The very nature of technology is that although it is often used to improve the human condition, these improvements usually come with a variety of consequences, some of which may be undesirable or unintended. Energy and power technology is a subject area for which there are many examples of this phenomenon. To study energy and power and recognize the myriad of ethical dilemmas that the use and development of such technologies present, it is necessary to understand some basic facts about energy and power systems.

One of the most important aspects of energy and power is the cost or economics associated with a particular source, and this is largely influenced by something called energy ratios. This concept hinders the development of many forms of alternative energy. An energy ratio is an expression of how much energy must be expended in order to capture and refine an energy source for use.

For instance, oil does not just flow out of the ground. It must be extracted, refined, transported, and dispensed all prior to use. How much energy must be expended to produce a gallon of gasoline that is ready for use at the pump? Now compare this to how much energy would be expended planting, maintaining, harvesting, and refining corn or sugarcane to produce that same amount of potential energy. The product would be about 1.2 gallons of ethyl alcohol for vehicular use. Then consider how much energy must be expended to produce a usable kilowatt-hour of electricity from a coal plant or a nuclear generating plant versus the energy that would be expended to produce that electricity via the use of photovoltaic cells, wind generators, or other alternative means of electricity production. Simply stated, an energy ratio is the proportion of energy in to energy out.

The less energy it takes on the input side to produce usable energy or power on the output side, the more economically favorable it is that a source will be developed and used. This is the essence of economics and energy, and one of the reasons that fossil fuels still dominate the energy landscape even with their associated environmental impacts.

Another important concept is that of efficiency. Efficiency is defined as the amount of energy or power output divided by the energy or power input and then multiplied by 100. Efficiency is expressed as a percentage, and greater efficiency means less wasted energy or power.

As an example, consider the efficiency of a small electric generator. The generator is capable of producing 2,000 watts of power. It consumes gasoline at a rate of .5 gallons per hour, and one gallon of gasoline is capable of producing approximately 100,000 BTUs of heat energy. Figure 7-2 illustrates how the efficiency of the generator would be calculated.

Figure 7-2. Calculating energy efficiency.

| | | | | |
|---|---|---|---|---|
| **Output** | 10 hp × 42.44 = 424 btu/min | **Input** | | 1666 btu/min |

$$\frac{424 \text{ btu/min}}{1666 \text{ btu/min}} \times 100 = 24\% \text{ efficiency}$$

When there are multiple conversions that must take place prior to a system yielding a desired form of power, then the total system efficiency is calculated by multiplying the efficiency of each conversion with the efficiency of the next conversion until all conversions are included in the calculation. Figure 7-3 represents a wind generator that is connected directly to the power grid. Note the total system efficiency.

Figure 7-3. Calculating total system efficiency.

| Blades capture wind at 57% efficiency. | Mechanical power is converted to electrical power at 90% efficiency. | Electrical power is transmitted to inverter at 98% efficiency. | Inverter conditions power for grid at 67% efficiency. |
|---|---|---|---|

Calculation: .57 × .90 × .98 × .67 = 33.7% total system efficiency

All methods of producing useful power have an impact on the environment. The most common forms of pollution from the consumption of fossil fuels include acid rain and the greenhouse effect. Acid rain is caused as a result of sulfur oxides, nitrogen oxides, and hydrocarbons mixing with raindrops and falling back down to earth, where they pollute forests, lakes, and streams. The greenhouse effect is most commonly associated with the phenomenon of global warming. It results from by-products of combustion that cloud our atmosphere, acting as a blanket around the earth. As light strikes the earth, it is reradiated as heat, but that heat cannot escape the earth's inner layers of atmosphere because of the buildup of greenhouse gases that hold the heat in. This leads to a warming of the earth's surface, changes in weather patterns that may adversely impact agricultural regions, and could eventually lead to flooding of coastal regions as polar ice melts.

Some forms of pollution, however, are considerably less subtle. Hydroelectric dams, for instance, produce comparably inexpensive electricity because there are no fuel costs. Hydroelectric plants also do not produce the pollutants that fossil fuel plants produce. The creation of a hydroelectric plant does result in a form of environmental pollution, forever modifying the ecosystem of the river by permanently changing the landscape and altering the flow conditions of the river. In a worst-case scenario, multiple dams can block the movements of migratory fish, thus preventing them from returning to their native breeding grounds. In some instances, this has contributed to the endangerment of entire species.

Even forms of power production such as wind generators that would be seemingly harmless to the environment do produce some forms of pollution. To generate large-scale power with wind as an energy source, many large wind generators are needed to create a wind farm. The generators cannot be placed very close to one another or the turbulence created by one wind generator will decrease the efficiency of the next generator. The result is that wind farms tend to occupy large expanses of land with massive wind generators that create noise and are viewed by many as unsightly.

It is an unfortunate reality that those energy sources that are the most environmentally friendly, such as wind, hydro, geothermal, and solar are also the least reliable. Most suitable sites for large-scale hydropower production in the United States have already been developed, and only a small percentage of those can be relied upon as base-load plants capable of producing a large and constant amount of electricity. The remaining hydro plants are at the mercy of the weather, capable of producing at full power when the river is adequately fed by natural rainfall, but often producing power at less than full capacity when natural rainfall is scarce. Similar to hydro power, solar power works best in those limited environments in which there are enough suitable yearly sun hours to make solar power production attractive, and wind power works best where a substantial breeze can be found most of the time. In the United States, electricity produced by solar power appears to have its best potential in the Southwest, and wind power would have its greatest potential along the coasts, on the plains in the Midwest, and in some mountainous inland regions.

Sustainable growth is another important factor related to energy and power. When discussing sustainable growth, the concept of doubling time is a phenomenon that has a tremendous impact on energy consumption patterns worldwide. As nations grow and strive to improve their standard of living, it is logical that in order to do so, more energy will be consumed. There is a correlation between the standard of living and the quality of energy and power available to a population. As the population grows and the standard of living increases, it is necessary to produce more power to meet the needs of the population, and the number of years it takes for power production to double becomes less. For instance, let's assume that a population of one million consumes 100 units of energy per year and that population is expanding and consuming more energy at a rate of about 6 percent yearly. The formula shown in Figure 7-4 allows us to calculate the doubling time before it would be necessary to produce 200 units per year to meet demand.

As can be seen, the doubling time on energy consumption would only be about 12 years at a 6 percent growth rate. This is a considerably shorter period of time than what might be expected. Yet, it is an unfortunate reality that for more nations to improve their standard of living, more energy will need to be consumed.

Figure 7-4. Calculating time to double energy consumption.

$$\frac{\text{Growth rate}}{71} = \text{Doubling time} \qquad \frac{6\%}{71} = 11.9 \text{ years}$$

It has been said that politics make for strange bedfellows. Throughout history, there is perhaps no better example of this observation than the politics of energy and power. Japan did not share all of Germany's political ideologies in the 1940s. Rather, a principal reason that Japan entered World War II was a desire to extend its reaches to lands that had natural resources, particularly energy resources that Japan did not possess. Likewise, the United States does not share political ideology with the governments of many members of the Organization of Petroleum Exporting Countries (OPEC), but it conducts a brisk trade business with them to support America's appetite for oil.

Importing oil allows the price of gasoline to remain low and ensures a plentiful supply, but it also places this nation at the mercy of other nations that control the flow of oil. Like any other commodity, when supply does not meet demand, the cost of that commodity increases. Twice in the 1970s, OPEC nations placed oil embargos on the United States, resulting in higher gas prices, gas shortages, long gas lines, gas rationing, and disruption to the economy. In the 1980s, America fought back with four cylinder automobiles and a host of energy conservation measures, driving the price of imported oil back down. After almost 10 years of relatively stable oil prices since 1980, America appears to have been lulled back into a comfortable mode of importing about half of its oil.

The politics of energy and power again became a prime concern to Americans in 1990 when Iraq invaded Kuwait. The result was the 1991 Gulf War, a war, which many contend, was primarily fought over the control of oil. However, oil prices remained stable and the Gulf War was over quickly, having little long-term effect on the level of oil imports to the United States.

## THE ENERGY GAME

*Description*

The intent of this activity is to help students learn more about energy and power as they experience some of the subtle ethical dilemmas associated with energy and power technologies. The game can be played individually or by up to six players at once. Players should read the preamble that leads up to this simulation activity so that they possess some minimal awareness of the issues and dilemmas surrounding the use of energy and power technologies.

*Educational Outcomes*

1. Students will be able to explain the common sources of energy and power.
2. Students will be encouraged to recognize the importance of preserving finite natural resources.
3. Students will develop awareness of interdependence of global communities as they participate in the Energy game.

*Materials List*

- One die
- One coin
- An object to use as a place marker on the game board
- Game board (master provided; see Figure 7-5)

- Red and blue game cards (masters provided; see Figures 7-6 and 7-7)
- Scorecard (master provided; see Figure 7-8)

*Game Scenario*

You have just been elected president of a tiny island with a population of about one million people. Every island generates its own electricity, and this is vital to enable a modern economy. Your job will be to continue to improve the standard of living for your people while respecting the environment and trying not to become too dependent on foreign energy sources. You will start out with $100 in working capital and 100 megawatts (MW) of excess generating capacity to protect the island from brownouts and peak demands. Decisions you make prior to the start of the game and at various points throughout the game will influence your results.

*Instructions and Game Rules*

1. Thoroughly shuffle the stack of red cards and blue cards. Then roll the die to determine which island nation you now lead. The number of the island rolled also determines the order of play.

2. Make your power generation preferences known by filling them in at the top of the scorecard so all players can see them. **You must maintain at least $50 in working capital AND 70 MW of excess generating capacity throughout the game or else you lose.**

3. Each player starts in the center of the game board and attempts to advance toward the right side of the board. Player 1 begins by flipping a coin to determine which stack of cards to chose from, either the red stack (heads) or the blue stack (tails).

4. Player 1 reads the appropriate card and then stays in place, advances to the right, or regresses to the left, depending upon the outcome of the card. **After moving your marker, you may purchase megawatts from other islands for cash or sell your own excess megawatts to other islands for cash.** If energy is bought or sold, the scorecards for all islands involved must be appropriately updated. You can only purchase or sell power while it is your turn. Play progresses in a clockwise fashion with Player 2 flipping the coin to take a turn and so on.

5. Any move to the left of the start area represents an increase in the standard of living for your island. **Deduct 6 MW of your reserve**

**power generating capacity for every square you move to the left unless told otherwise.**

6. The game is over when one player reaches the far right of the game board (total energy and power enlightenment), or when **all** other players have met one of the following conditions:
   a. Been voted out of power by reaching the far left side of the game board
   b. Failed to maintain minimum working capital of $50
   c. Failed to generate a minimum of 70 MW of electricity to sustain the economy

Alternative versions of the game could be used in which participants either play for a set period of time or play for a set number of turns, and then progress on the game board is compared to determine a winner. When used as a classroom experience, it is recommended that the topic be first introduced, followed by students reading a copy of the introductory materials in this section to establish a minimal knowledge base. Play the game for 20 minutes maximum and then move on to the concluding questions that form the basis for a discussion about energy, power, and ethics.

*Reflection Questions*

After playing the Energy game, have the students consider the following questions and formulate responses for each:

1. Are there strategies for managing energy consumption or is it based on luck?
2. Is there one form of energy that can solve all human wants and needs?
3. Nuclear power is known to produce inexpensive electricity, but has the final bill really been paid with regard to the storage of nuclear waste?
4. Is it possible to increase the standard of living while increasing generating capacity and protecting the environment?
5. Is importing energy from foreign countries beneficial or detrimental?
6. Are there forms of energy available to us that are truly nonpolluting?
7. To what extent should one government interfere with another to protect its own energy policy?

Figure 7-5. Game board for the Energy Game.

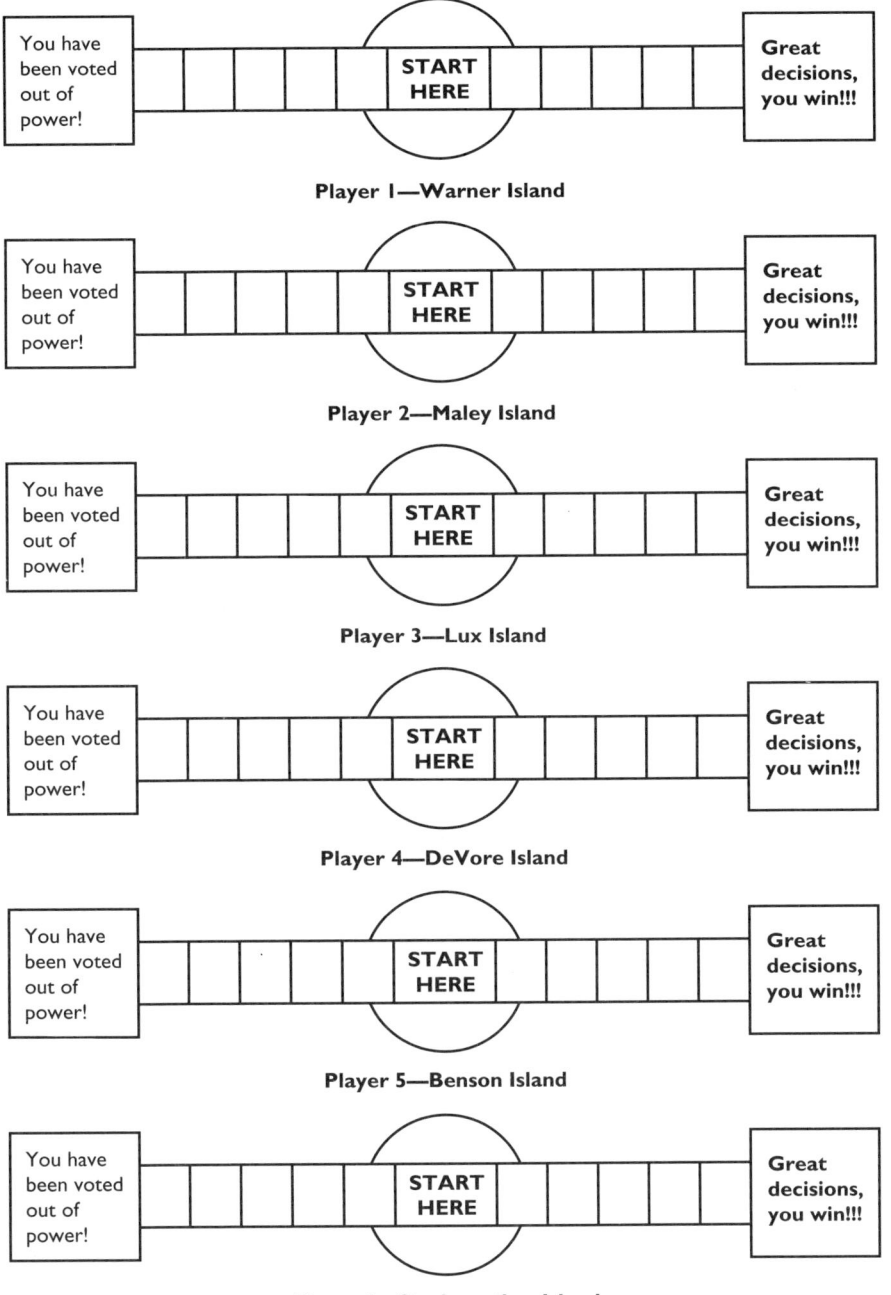

Figure 7-6. Red cards (heads on coin toss) for the Energy Game.

| | |
|---|---|
| The economy sours. Deduct $5 of your working capital and go back one space. | A coal miner's strike reduces the amount of energy to all islands by 5%. Deduct 5 MW from your generating capacity and go back one space. |
| Talk of war scares investors. Deduct $10 of your working capital and go back one space. | Protesters outside the nuclear plant cause the plant to shut down. If you generate with nuclear power, deduct $10 of your operating capital to purchase alternate power OR reduce production by 10 MW and go back one space. |
| A shortage of fossil fuels drives demand up. Deduct $7 of your working capital and go back one space. | Acid rain is ruining the environment of Maley and Warner Islands. Deduct $10 of your working capital if you are president of either one of these islands and you are generating electricity at more than 40% capacity with coal. Then go back one space. |
| You will have to pump more capital into pollution control. Deduct $6 of your working capital and go back one space. | A terrorist threatens to blow up a nuclear plant on your island. Deduct $6 of your working capital to pay for increased security, if you generate with nuclear power. Then go back one space. |
| Residents of your island vote to close down the nuclear plant because of environmental concerns. If you generate with nuclear power, deduct the megawatts generated from your nuclear plant and go back one space. | If you are president of a nation that generates at over 50% capacity with a combination of coal or oil, then donate 3% of your working capital to cancer research. Stay put where you are. |

Figure 7-6. Red cards (heads on coin toss) for the Energy Game (continued).

| | |
|---|---|
| Political ideologies make trade between other islands with Warner Island and Maley Island impossible. If you bought megawatts from Maley or Warner Island, deduct 10 MW of your generating capacity and go back one space. | Any country generating at over 10% with hydro is being sued by environmentalists who are concerned that hydro plants ruin the ecology of the river. If you generate with more than 10% hydro, deduct $5 of your working capital for legal fees and go back one space. |
| A typhoon has swept through the region severely damaging DeVore Island and Lux Island. If you are president of either of these islands, divert $10 of your working capital to disaster relief, reduce production by 5 MW, and go back one space. | You are caught buying illegal oil. The United Nations has issued a hefty fine costing you $10 of your working capital and 5 MW of power. Go back two spaces! |
| A drought has struck Benson and Starkweather Islands. If you are president of one of these islands and generate with hydro at more than 20% capacity, deduct $10 of your working capital to purchase power, reduce generating capacity by 5 MW, and go back one space. | Storing nuclear waste has become a global problem. Contribute $7 of your working capital to a fund to find a long-term solution for the safe storage of nuclear waste. Stay put where you are. |
| DeVore Island will not sell you any more enriched uranium. If you chose to generate with nuclear power, then the cost has just gone up. Deduct $5 from your working capital OR subtract 5 MW from your power production. Stay put where you are. | You failed to meet your obligation to the United Nations for alternative energy research. Now, you must pay $7 of your working capital to the fund to remain on good terms with other nations. Stay put where you are. |
| Solar research on your island turns out to be a flop. If you use solar power as a generating means for the future, then deduct $5 from your working capital and go back one space. | If you are president of an island that generates at over 50% capacity with a combination of coal and oil, then donate $5 of your working capital to cancer research. Stay put where you are. |

*Ethics and the Study of the Designed World*

Figure 7-7. Blue cards (tails on coin toss) for the Energy Game.

| | |
|---|---|
| Conservation measures decrease energy demand by 5% on your island. Add $5 to your working capital. Deduct 6 MW and advance one space. | You invest in tidal power as a potential energy source for the future. Deduct $10 from your working capital, deduct 6 MW, and then advance one space. |
| An alternative energy breakthrough occurs on your island. No megawatt hour reduction is required. Advance two spaces! | New turbine configurations allow your island to generate an additional 10 MW of power. Add 10 MW to your total and advance one space. |
| A glut of oil on the market drives prices down to their lowest level in five years. Add $5 to your working capital. Then deduct 6 MW and advance one space. | Cogeneration will allow a private-sector company to purchase excess steam from your coal plants. If you generate with coal at 30% or greater, add $10 to your working capital and advance one space. No megawatt hour reduction is required! |
| You invest in geothermal power as a potential energy source for the future. Deduct $10 from your working capital. Then deduct 6 MW and advance one space. | A mild winter results in energy costs that are lower than expected. Add $10 to your working capital, deduct 6 MW, and then advance one space. |
| The United Nations is interested in the development of alternative energy sources for third-world nations. If you generate with 10% alternative energy, deduct 6 MW of power and advance one space. Otherwise, stay put. | Recent political partnerships allow you to purchase excess electricity at less than market value. Add 10 MW to your total and advance one space. |

Figure 7-7. Blue cards (tails on coin toss) for the Energy Game (continued).

| | |
|---|---|
| You invest in ocean thermal energy conversion as a potential energy source for the future. Deduct $10 from your working capital. Then deduct 6 MW and advance one space. | Your electric vehicle program is taking off, but it necessitates more generating capacity. Deduct $10 from your working capital. Then deduct 6 MW and advance one space. |
| You invest in wind energy as a potential energy source. Deduct $10 from your working capital. Then deduct 6 MW and advance one space. | Increases in the efficiency of climate control systems are beginning to have an impact on energy consumption. Add 10 MW to your total and advance one space. |
| An exceptionally rainy season allows you to generate excess capacity from your hydro stations. If you generate with hydro, add $5 to your working capital and 5 MW to your power production. Then advance one space. | Governments praise your fossil fuel emissions improvements, but the control devices for emissions are costly and they consume an additional 3 MW. Deduct 3 MW and $5. Then advance one space. |
| Breakthroughs in alternative energy research allow you to generate more electricity than expected. Add 10 MW and advance one space. | The *Union of Concerned Scientists* rates your method of storing nuclear waste as the best of all the islands. Advance one space. No megawatt hour reduction is required! |
| Scientists find a deep pocket of coal on your island that has been previously undetected. Add $7 to your working capital and advance one space. No megawatt hour reduction is required! | New applications for the by-products of coal power reduce disposal costs. If you generate more than 30% power with coal, then advance one space. No megawatt hour reduction is required! |

*Ethics and the Study of the Designed World*

Figure 7-8. Scorecard for the Energy Game.

The Energy Game Scorecard  _____ **Island**

**Power Generation Preferences**

| | | | | | | | |
|---|---|---|---|---|---|---|---|
| Coal | _____ (50% max) | | | $___ +/– = _____ | | $___ +/– = _____ |
| Nuclear | _____ (50% max) | | | $___ +/– = _____ | | $___ +/– = _____ |
| Hydro | _____ (30% max) | 100% MAXIMUM TOTAL | | $___ +/– = _____ | | $___ +/– = _____ |
| Oil | _____ (25% max) | | | $___ +/– = _____ | | $___ +/– = _____ |
| Solar/Alt. | _____ (10% max) | | | $___ +/– = _____ | | $___ +/– = _____ |

$___ +/– = _____    $___ +/– = _____

| Capital | Megawatts |
|---|---|
| $100 +/– = _____ | $100 +/– = _____ |
| $___ +/– = _____ | $___ +/– = _____ |
| $___ +/– = _____ | $___ +/– = _____ |
| $___ +/– = _____ | $___ +/– = _____ |
| $___ +/– = _____ | $___ +/– = _____ |
| $___ +/– = _____ | $___ +/– = _____ |
| $___ +/– = _____ | $___ +/– = _____ |
| $___ +/– = _____ | $___ +/– = _____ |
| $___ +/– = _____ | $___ +/– = _____ |
| $___ +/– = _____ | $___ +/– = _____ |
| $___ +/– = _____ | $___ +/– = _____ |

$___ +/– = _____    $___ +/– = _____
$___ +/– = _____    $___ +/– = _____
$___ +/– = _____    $___ +/– = _____
$___ +/– = _____    $___ +/– = _____
$___ +/– = _____    $___ +/– = _____
$___ +/– = _____    $___ +/– = _____
$___ +/– = _____    $___ +/– = _____
$___ +/– = _____    $___ +/– = _____
$___ +/– = _____    $___ +/– = _____

8. To what extent should environmental concerns influence the need for increased power production and continued exploitation of energy resources?
9. To what extent do developed nations have an obligation to share energy and power technology with less-developed nations?
10. If providing nuclear technology would improve the standard of living for third-world nations, should that technology be provided to them?

*Instructor Notes for Reflection Questions*

1. Some things such as natural disasters cannot be predicted, but there is some strategy to the game that tends to yield a greater potential for success. The strategy has to do with diversifying energy resources so as not to rely too much on any one resource.
2. At present, there is no one form of energy that can fulfill all human needs. Some energy sources, such as nuclear and hydro, can only be used to generate electricity, but transportation accounts for approximately 27 percent of all energy consumption in the United States, and the vast majority of all transportation cannot be powered by electricity.
3. There is presently no long-term storage facility available for the maintenance of high-level radioactive waste, although a Yucca Mountain site on federal land in Nevada appears to be close to approval. Most of the radioactive waste generated by this nation's approximately 104 operable nuclear plants sits on-site in storage pools at the various power plants. Although utility companies continue to pour funding into a holding account to deal with the long-term storage of nuclear waste, the final bill on the cost to store the nuclear waste has yet to be paid.
4. It is possible to increase the standard of living while protecting the environment. The answer lies in the form of technological breakthroughs, such as scrubber systems and other environmental pollution controls that allow power to be generated with less harmful by-products seeping into the environment. Creative solutions, such as using fly-ash—a common by-product of coal generation—as road grit for traction in snow, have reduced the need to landfill a formerly useless waste product.

5. Importing energy is a double-edged sword. It allows the United States to be a good trading partner and to extend some influence over the foreign governments that are trade partners, but it also leaves the United States vulnerable to external forces on the economy and sometimes requires partnering with governments that share vastly different ideologies and political structures.

6. It has been said that if a 140 square mile area of Arizona were covered with photovoltaic cells, they could provide all of the electricity needs for the United States on a yearly basis. Technologies that make use of the natural environment and that do not require combustion to produce energy tend to be much safer for the environment. Advancement in technologies such as fish ladders and elevators now allow the power companies to generate hydropower while living in relative harmony with migratory fish.

7. Questions 7–10 are significantly open-ended to the point that there is no one correct answer, but they are perhaps the most interesting of all of the questions in terms of ethics. Any sharing of energy technology would certainly have to be balanced by a concern for safety, national security, and the environment.

# INFORMATION AND COMMUNICATION TECHNOLOGIES

*Mark Sanders*

The premise of this yearbook is that preparing "technologically literate" citizens makes it imperative for technology teachers to educate students not only about the nuts and bolts of various technologies—in this case, communication systems—but also about the sociocultural issues and impacts relating to those technologies. Perhaps more than any other technological arena, information and communication technologies are riddled with critically important issues that go to the heart of our democratic society.

Many of these issues are politically charged in nature. Addressing issues such as these in a technology education program, for example in a communication technology course, is a radical departure from our historical practice of teaching only the mechanics of various communication technologies. Teaching the technical processes necessary to produce digital audio and video, for example, is one thing. Debating the associated ethical questions, such as the right and wrong of global theft of copyrighted music and videos, is quite another.

One cannot address the ethics of communication technology without coming to grips with many such complex and potentially controversial issues. The transition from analog to digital communication systems that occurred in the last half of the twentieth century and the resulting convergence of print, audio, video, and data communication systems into one omnibus communication system have completely transformed our culture. The "information age" has become the networked digital information age, and the simple fact that text, audio, video, and data are all now part of one globally accessible digital medium has forever changed the landscape, our culture, and our world. In short, our global network of wired and wireless technologies has opened up an entirely new Pandora's box of "new age" ethical issues.

The activities suggested in the following pages should not be undertaken lightly. Technology educators who go down this path will potentially, perhaps likely, thrust themselves into a dialogue with students, administrators, parents, and politicians in their local community. The various ethical issues that emerge from this path have sometimes been discussed in the social studies curriculum, but those dialogues usually remain relatively private, constrained by the walls of the social studies classroom. The following activities will necessarily move the conversation into the public realm, so communication technology teachers should proceed with relative caution.

As technology educators begin to address these questions of ethics and values, they should be aware that social studies educators are also increasingly engaged in these same discussions. The National Standards for Social Studies Teachers (National Council for the Social Studies, 1997) includes explicit language regarding the importance of the study of technology within social studies. Known as "Thematic Standard #8, Science, Technology, and Society," this standard encourages social studies teachers to incorporate the study of technology in new and important ways. Specifically, part of that standard states:

> Tracing the impact of science and technology in such areas of human endeavor as agriculture, manufacturing, the production and distribution of goods and services, the use of energy, communication, transportation, information processing, medicine and health care, and warfare enables learners to understand both the way science and

technology have influenced and have been influenced by individuals, societies, and cultures.

For this reason, it behooves technology educators who engage in the teaching of ethics and values, such as those suggested herein, to coordinate these efforts with the social studies teachers in their respective schools. Because social studies teachers will likely be discussing some of these same communication technology issues in their classes, technology teachers should take advantage of an excellent opportunity to forge new connections between the two subject areas. This is an interdisciplinary opportunity that should not go unexplored.

The following technology education activities are suggested as a means of engaging students in the study of some of the ethical issues and questions associated with new information and communication technologies.

## ETHICS AND VALUES DEBATE

### *Description*

Students debate ethical issues related to information and communication technologies. Students have opportunities to learn about differing perspectives on choices they might reasonably be expected to face.

Technology teachers should consider working with a social studies teacher on this particular activity. One approach might be to hold the debate jointly with a social studies class. Alternatively, the debate might be a social studies class activity, during which the technology education students are seated as an audience for the debate. In that way, the technology education students would benefit from hearing the arguments and ideas forwarded during the debate, thereby providing them good background information as they begin the "Promotional Campaign" activity described in the following section.

### *Educational Outcomes*

1. Students will identify ethical issues relevant to information and communication technology.
2. Students will practice oral communication skills as a part of a debate activity.

3. Using a debate format, students will become familiar with positive and negative aspects of choices related to information and communication technologies.

## Materials List

- Copy of assigned debate topic for distribution to students
- Reference materials relevant to assigned debate topic
- Peer evaluation form (teacher developed) for critique of debate team presentations

## Instructions

1. Divide the class into suitable debate teams (for example, six students per topic selected in the following step, with three students on each side of the issue).
2. Assign one of the following topics/questions to each of the student debate teams. [Note: The following issues are stated as questions; however, these might be posed as "resolutions" instead. For example, the first issue might be stated as: Whereas . . . Spam on the Internet should not be regulated by governmental agencies.]
   a. To what extent should "spam" be allowed to proliferate unchecked on the Internet?
   b. Should environmental impacts of various communication technologies (for example, cell phone and computer disposal, satellite dish or cell phone tower location, and so on) be more closely regulated by governmental agencies?
   c. In 1995, because copyright protection was about to expire on several famous characters copyrighted by the Walt Disney company, the Disney corporation successfully lobbied the U.S. government to rewrite the copyright laws to extend the period for an additional 20 years. Thus, in 2015, this request will likely be issued once again. In light of this probability, should copyright protection be extended beyond its current limits?
   d. To what extent should "sampling"—the practice by which an individual makes a digital copy of a segment of commercial music, alters this sampled audio with digital tools in some way, and then uses or sells this new derivative for profit—be regulated by government?

## Ethics and the Study of the Designed World

  e. Cybercrime legislation: Select a current issue of this nature and debate the pros and cons.

  f. Right to privacy versus national security and freedom of information: Select a current issue of this nature and debate the pros and cons.

  g. Right to free speech versus pornography on the Web: Select a current issue of this nature and debate the pros and cons.

3. Review debate procedures with students. See, for example, resources found at the National Federation Interscholastic Speech and Debate Association (http://www.nfhs.org/nfisda.htm), National Forensic League (http://debate.uvm.edu/nfl.html), and Urban Debate Leagues (http://debate.uvm.edu/udl/udl.html) Web sites.

4. Provide sufficient time and resources for students to conduct research. Encourage students to use such resources as:

  a. Edupage Listserv (http://www.educause.edu/pub/edupage/edupage.html) and the Edupage Archives (http://listserv.educause.edu/archives/edupage.html). Edupage is a free e-mail service that summarizes developments in information technology. It is issued three times a week to subscribers. Among these brief news abstracts are many that address ethical issues.

  b. Web search engines such as Google, Yahoo!, HotBot, and so on.

  c. Electronic encyclopedias and reference materials, such as Encarta, World Book, Grolier's, and so on.

  d. Online books and articles, accessible from local, public, and/or university libraries.

5. Develop a peer evaluation that students may use to provide peer feedback.

6. Moderate the debate. Consider inviting other students, teachers, administrators, and parents to this forum.

# PROMOTIONAL CAMPAIGN

*Description*

Students develop a promotional campaign for their school or community that promotes better understanding of some of the issues and questions relating to new information and communication technologies.

This promotional campaign activity is perhaps best suited to a high school or university communication technology class. Structuring a Communication Technology class around the development of a promotional campaign has worked very well for me in my university-level Communication Technology course for more than a dozen years. This activity should engage the entire class of students throughout most or all of a semester. Consider grouping students in teams; teams of three students is a good team size for these activities. The following list of "deliverables" may be altered to suit your situation. The idea is to customize the list so it works in your situation. Review the procedures for each of the selected "deliverables" with students at the beginning of the semester.

Obviously, this activity must be preceded by or complemented with technical instruction in the various areas represented in the activity. To a large extent, the number of "deliverables" you choose to include will depend upon both your facilities and the prerequisite knowledge your students bring to this course. If this is their first technical course in this area, you'll want to select only a couple or more of these deliverables. In situations in which students have rich technical backgrounds in these communication technologies, you may decide to select most of the deliverables.

If students are working in teams, develop a peer evaluation they may use to critique their teammates both at midterm and at the end of the semester. Consider using the debate on ethics and values described in this section early in the semester to provide students with a good background for their work with the promotional campaign. Be certain to encourage their use of the types of resources identified in the Materials List describing the ethics and values debate.

## Educational Outcomes

1. Students will identify ethical issues relevant to environmental impacts, intellectual property rights, free speech, and national security that impact uses of information and communication technologies.
2. Students will develop or enhance skills using electronic or print media to effectively communicate messages related to ethics in a technological world.

## Materials List

- Copy of the design brief for the promotional campaign
- Reference materials relevant to information and communication technology ethics
- Software for development of Web materials
- Access to Web space to host online materials for the promotional campaign
- Screen printing or inkjet appliqué materials for T-shirts
- Digital photography equipment
- Digital video camcorder
- Digital video editing software and related hardware
- Color printing equipment and paper

## Instructions

### Design Brief for the Promotional Campaign

Develop a promotional campaign for your school and community that promotes better understanding of the many issues and questions related to new information and communication technologies. Issues to consider addressing in this promotional campaign include:

1. Environmental impacts
2. Plagiarism and copyright infringements with regard to text, images, music, and video
3. The first Amendment to the U.S. Constitution and the balance among issues such as freedom of speech, spam, censorship, and pornography

4. The fourth Amendment to the U.S. Constitution and the balance between national security—for example, electronic surveillance via the Internet—and the right to privacy

Consider the audience for your promotional campaign to be your school community: students, teachers, administrators, and parents.

### Deliverables and Procedures for Students

1. Design and develop a Web site with sections addressing the major communication technology ethical issues and questions, such as those noted previously.
   a. Research these topics using resources such as those identified in Activity #1.
   b. Create the content for the Web site in the form of written reports on various ethical issues and questions—such as those noted previously—developed by student teams. These reports should include the following sections:
      i) Introduction
      ii) Description of the Issue. This section should describe the issue, being careful to cite all information sources used.
      iii) Questions to Consider: This section might pose questions that would engage the audience.
      iv) References. Information sources used (Web sites, online data, and so on) should be cited properly according to an accepted style guide (for example, Publication Manual of the American Psychological Association).
   c. Convert the reports to HTML pages
   d. As appropriate, add copyright-free graphics, music, video, and so forth to your HTML pages to make them more appealing and informative for your audience.
   e. Add links on your Web pages to appropriate resources, related research, and so on.
   f. Consider the following design recommendations as you design your Web site:

- i) Design with the principles of contrast, repetition, alignment, and proximity in mind (Williams and Tollet, 2000).
- ii) Small files (as measured in KB) load quickly on Web pages and are much preferred to large files.
- iii) Use only the 216 "Web-safe" colors.
- iv) Limit the line length of text to about 12–15 words/line with fixed-width invisible tables.
- v) Contrast between text and background is critical for readability. Black text on a white background provides optimal contrast.
- vi) Navigation should appear consistent throughout the site (for example, a menu that has the same general appearance and location on every page).
- vii) Media (audio and video) should be streamed or else less than 1 MB in file size.
- viii) Evaluate the appearance of your Web pages in different browsers on different computer platforms.
- g. Incorporate Web pages, PDFs, and so on, from each of the following items into this Web site.

2. Design and screen-print a one- or two-color T-shirt (depending upon equipment availability) with a logo or slogan that raises awareness about one or more ethical issues relating to information and communication technologies.
3. Design and produce a color brochure for the student body, teachers, administrators, and parents that raises awareness about the ethical issues.
4. Shoot a digital photo-essay that addresses one or more ethical issues related to information and communication technologies.
5. Design, using a storyboard and script, and produce a digital video public service announcement (PSA) about one or more ethical issues related to information and communication technologies, which might be aired on your school video network or on your local public television station.

*DeMiranda, Benson, Litowitz, Sanders, Seymour, Wescott, Womble, and Williams*

# WEBQUEST: ETHICS AND TECHNOLOGY IN THE CLASSROOM

*Description*

Webquests are activities that allow students to access various materials on the World Wide Web related to a specific topic. Guidance and instructions for the activity described here may be found on the Web using the URL that follows. *Ethics and Technology in the Classroom* (http://www.tandl.vt.edu/technology-ethicswebquest/scenarios.htm) provides students with five different "ethical issue" scenarios:

1. Copyright and plagiarism
2. Be a critical user of the Internet
3. Acceptable use policies in the classroom
4. Online safety and ethics in the classroom and at home
5. Other related issues

*Educational Outcomes*

1. Students will use online resources to investigate ethical issues related to information and communication technologies.
2. Students will enhance their skills in using the World Wide Web for research.

*Materials List*

- Computer connected to the Internet
- Web browser software

*Instructions*

Students should launch their Web browser software and access the Web site titled *Ethics and Technology in the Classroom* located at http://www.tandl.vt.edu/technology-ethicswebquest/scenarios.htm. Here, they will find hyperlinks to the various scenarios that are a part of this activity.

Each scenario begins with guiding questions and then directs students to resources on the Web that are intended to assist them in formulating answers. After students have been through these five scenarios, they should be instructed to develop a Web site and create links to the following products.
1. Word-processed answers to the "guiding questions" for each scenario
2. An Internet user's guide for students that addresses copyright issues, searching, and evaluating sources
3. An Acceptable Use Policy (AUP) for their classroom
4. A PowerPoint presentation and classroom poster that addresses safety issues on the Internet
5. A list of the top 10 issues teachers face in the classroom regarding ethics and technology. This will be based on all five scenarios.

These scenarios, guiding questions, and products may be used to supplement the Web site outlined in the promotional campaign activity described previously. Students can be directed to complete just one, several, or perhaps all of the scenarios and products before they undertake the promotional campaign activities.

# TRANSPORTATION TECHNOLOGIES
*Myra N. Womble and Stephanie Williams*

## HYBRID ELECTRIC VEHICLE RESEARCH

*Description*

The purpose of this activity is for learners to use findings from research to enhance their knowledge of hybrid electric vehicle (HEV) technology and related critical issues. Learners conduct research using the Internet and other resources, such as local businesses and industries, and then engage in learning activities that allow practice of writing skills, desktop design skills, communication skills, and collaborative learning.

*Educational Outcomes*

1. Students will use online resources to investigate ethical issues related to transportation technologies.

2. Students will develop enhanced skills using research to support writing, desktop design, communication, and collaborative learning.

## Materials List

- Computer connected to the Internet
- Web browser software
- Word processing or desktop publishing software for development of brochure
- Paper and printing materials for brochure

## Instructions

This activity is designed to be completed in two parts. Part 1 involves use of the Internet and other sources to collect information about hybrid cars. Part 2 involves design and production of a hybrid car brochure using information collected when completing Part 1. The information that follows can be used to provide an introduction to the HEV research activities.

Not since 1916, when Woods invented a hybrid car that had both an internal combustion engine and an electric motor, have Americans given so much attention to alternative means of propulsion for their passenger cars (Bellis, n.d.). Some supporters of today's hybrid cars believe the result of this renewed attention will be the same as was seen with electric cars in 1899 and 1900—with HEVs eventually outselling all other types of cars. HEVs were initially purchased mostly by environmentalists, technology enthusiasts, and celebrities. However, recent gulf conflicts, a troubled economy, and increasing gasoline prices have made hybrid cars attractive to a widening array of consumers.

HEVs are steadily gaining in popularity, and, to some, have become a status symbol. One HEV, the Toyota Prius, which runs on petrol and electric power, was made available in 2000 and sold about 3,500 over a two-year period (Baber, 2002). Now, HEVs are being sold at a rate of almost 5,000 per month (Freeman, 2003). Encouraged by consumers, government, and profit, manufacturers are developing HEVs that will address issues of air quality and dependence on foreign oil, while still attempting to provide the kinds of vehicles Americans want and need (Parker, 2003). The following activity provides learners an opportunity to learn more about HEVs by conducting face-to-face interviews with local experts and conducting Internet research to determine the answers to the following Learner Questions.

### Part 1 Activities

1. Divide the class into groups of three to four learners and assign the 10 following questions to each group.
2. Instruct each group to conduct Internet research to determine the answers to each of the 10 questions. Online resources might include free sites as well as subscription sites for newspapers and magazines.
3. Instruct each group to select a model of hybrid car sold at a nearby dealership. They should then arrange an interview with a dealer or salesperson and take notes on their answers to each of the 10 questions as they relate to the hybrid car selected. In preparation for this activity, the instructor should identify local experts and Internet sites. This activity can be modified to address different cities and regions worldwide.
4. Instruct learners to use the information they have collected to develop a class presentation about hybrid cars. Each group should present its research to the class. Afterward, engage learners in a discussion about hybrid cars that allows comparison and contrast of ideas and information discovered by each group.

### Learner Questions

1. How does the technology of an HEV work?
2. What goes on under the hood to give the consumer more miles per gallon than is provided by the standard automobile?
3. How does an HEV pollute less, other than just getting better gas mileage?
4. What special instructions do consumers need about how to drive an HEV for maximum efficiency?
5. How do HEVs compare to standard automobiles in terms of comfort, appearance, and interior features?
6. What HEVs are available now, what are the most popular HEVs, and what are the expectations for the future?
7. Where can up-to-date information be located about future technologies and HEVs being developed for the market?
8. What would be involved in converting a standard automobile into an HEV?

9. How long do HEV batteries last and how much do they cost? What are the projections for change in life and cost of these batteries?
10. What research has been done to show the effects of HEVs on the environment? What are future projections for benefits to the environment?

**Part 2 Activities**

Using the information collected when answering the research questions in Part I, design a brochure that depicts information about hybrid cars of interest to consumers who are considering purchasing a hybrid car. Follow these basic guidelines when developing your brochure:

1. Use a 11" × 8.5" page size, trifold, two sides, black and white or color depending on printing technology available.
2. Follow basic principles of design.
3. Include a lead story and organize information into additional stories.
4. Include graphics and photos.
5. Include group identification and other sections as appropriate. Sections might include: Description of HEV technology, Advantages and Disadvantages of HEVs, Use and Care of HEVs, Popular HEVs, Future of HEVs, Where to Find Information About HEVs, HEV Conversion, Costs, and Environmental and Conservation Issues.

*Assessment Tools*

These activities lend themselves nicely to individual rubrics that can be collaboratively authored by the learners and the facilitator. Design of rubrics might be considered as a third activity, perhaps attempted by each of the learner groups initially formed, then blended for final rubrics.

## PUBLIC OPINION POLL

*Description*

The purpose of this activity is for learners to use findings from research to further enhance their knowledge of sports utility vehicle (SUV) technology and related critical issues. Learners design a survey to investigate consumers' perceptions and attitudes toward the SUV. Developing the survey instrument and collecting the data reinforces learners' research skills, writing skills, and communication skills, as well as engages them in a collaborative learning activity.

In preparing learners to become active members of society, we expose them to a variety of experiences that involve the cognitive, psychomotor, and affective domains of learning. Teachers have the opportunities to inform and discuss with learners ideas and situations that impact their lives. Transportation affects all of us daily from the simple, such as buying groceries, to the complex, such as terrorism.

We are most familiar with land transportation, but we cannot overlook air, space, and water transportation and their effects on how society functions. When examining how we understand and use transportation technology, simple and complex ethical issues arise. Today, transportation technology is easily accessible and is used routinely by the vast majority of society. Judgments we make about how we use this technology affects our surroundings, including the people we come in contact with and the air we breathe. The following activity encourages learners to reflect on their current and future decisions as consumers in a capitalistic society.

## *Educational Outcomes*

1. Students will conduct a survey to determine people's perceptions of a common mode of transportation.
2. Students will develop transportation concepts and criteria for selecting a personal vehicle.

## *Materials List*

- Computer connected to the Internet
- Web browser software
- Word processing or desktop publishing software for development of brochure
- Paper and printing materials for brochure

## *Instructions*

This activity is designed to be completed in two parts. Part 1 involves use of the Internet and other resources for collecting information to develop a survey instrument that focuses on consumer attitudes about SUVs in terms of environmental issues, costs, fuel performance, safety, and other concerns. Part 2 involves development of the survey instrument using information collected when completing Part 1, as well as data collection, data analysis, and reporting of findings.

### Part 1 Activities

1. Divide learners into groups with three to four persons in each group.
2. Instruct learner groups to develop questions that would be useful in obtaining information from consumers about their preferences, likes, dislikes, and knowledge about SUVs. Resource materials used during this activity might include online and print materials produced by automobile manufacturers that describe the many features now available on SUVs.
3. Provide a brief introduction to survey research and discuss Likert-type survey instruments with learners. Share examples of survey instruments and results from surveys with learners to help them begin forming their own survey questions and thinking about how to write results.

### Part 2 Activities

1. Bring the entire class back together and provide each small group with an opportunity to present their survey questions. Record the questions using a flip chart, marker board, or computer with projector. Place tally marks beside repeated questions.
2. After all groups have presented their work, discuss the questions that have been developed and reach consensus on 15–20 items to be included on a survey to be used by the entire class.
3. Have learners conduct a pilot-test survey with friends and family. Learners are to note respondents' problems with the survey instrument. Instruct learners to share and use this feedback to make changes to the survey instrument.
4. After the survey instrument is completed, instruct each learner to be responsible for collecting data on 10 completed surveys.

5. Instruct learners to return to their original groups to compile and tabulate the survey data. Learners should use percentages and frequencies to report the data. Spreadsheet software can be used to assist with this process if available.
6. Provide an opportunity for each group to report findings in a presentation.
7. Instruct learners to write a reflective analysis of their participation, concepts, and information learned and how the information obtained could be used.

### Learner Questions

1. What are the advantages and disadvantages of owning an SUV?
2. How do SUVs compare to standard automobiles in terms of comfort, appearance, and interior features?
3. What SUVs are available now, what are the most popular SUVs, and what are the expectations for the future of the SUV?
4. Where can up-to-date information be located about future automotive technologies and SUVs being developed for the market?
5. What research has been done to show the effects of SUVs on the environment? What are future projections about the impact of SUVs on the environment?
6. Describe the SUV tax break to business owners.
7. What is the federal regulation for bumpers? Why does the federal bumper standard not apply to SUVs? What changes in federal bumper standards would you support?
8. The National Highway Traffic Safety Administration research has shown that taller, narrower vehicles, such as SUVs, are more likely than lower, wider vehicles, such as passenger cars, to tip and roll over once they leave the roadway. What is the Rollover Resistance Rating for four different SUV models?
9. Is there currently a decline or rise in consumer interest in buying SUVs?
10. Some say there are few legitimate reasons to purchase an SUV. What are some legitimate reasons? What are some myths about SUVs that might encourage consumers to buy them?

# MANUFACTURING TECHNOLOGIES

*Richard D. Seymour and Michael A. DeMiranda*

Manufacturing technology involves using raw materials to produce items that satisfy our wants and needs. That simple definition fails to link the term "manufacturing" to ethics and responsibility. Or does it? For instance, what if the primary resources are found in an old growth forest that is off-limits to logging? What if the raw materials must be shipped around the globe, endangering coastlines or sensitive habitats? What if excessive waste from the production cycle threatens the air and land near the factory? What if the product satisfies all human expectation and is used all over the globe, yet is not conveniently disposable?

The modern factory floor is already a complex environment. But today, production firms must deal with numerous internal and external challenges beyond material and operational issues. Examples of integrity and ethics include (Gorman et al., 2000; Poe, 1983; Sobek, Liker, and Ward, 1998):

- Product safety and reliability
- Quality versus cost decisions
- Design for planned obsolescence
- Honesty in planning and operational tasks
- Attitudes and actions of employees
- Deceit during the bidding process
- Respect for patent and trademark rights
- Adherence to environmental guidelines
- Fair labor practices (both internal and among foreign suppliers)
- Responsibility in packaging
- Disposal of waste

It would be difficult to address each of these topics in a short section of one book. Yet, challenging issues arise daily in manufacturing enterprises, both on the production floor and in the front office. Although most of the issues involve a clear "right" versus "wrong" connotation, there are also areas of gray in many of the situations. Top management, supervisors, and workers must be prepared to recognize the appropriate course of action in dealing with all types of issues. As Mulgan (1997, p. 156) reminds us "... we are programmed to think morally."

## *Manufacturing Education*

Perhaps due to past fiascoes, the issue of ethics in manufacturing and engineering has been increasingly addressed in schools of higher education. A number of management and technical programs have turned their focus to transforming their designers, engineers, and managers "into ethical practitioners who will reflect on the consequences of their designs and discoveries" (Gorman et al. 2000, p. 463). The goal is to better prepare manufacturers to deal with complex challenges.

Business and management schools lead the way on addressing ethics in production, with the engineering community close behind. Most professional associations in technical and technological fields have also completed a code of ethics for their membership. For example, Pond (1993, p. 66) notes how the Code of Ethics for the Instrument Society of America "helps technicians and technologists to place their work in ethical perspective." The objective is to encourage the highest level of conduct while avoiding embarrassing and illegal actions.

Technology educators have also developed curricula and units of instruction that address ethics. The national content standards have helped by establishing a link between technology and appropriate action (for example, Standards 1–7 and 13). These themes now routinely appear in secondary and undergraduate programs. Much of the early focus was focused on environmental issues, but corporate and personnel responsibility are now included in the discussion. This book is another example of drawing attention to integrity and ethical decision making.

One way to introduce ethics in manufacturing classrooms is through case studies based on either real events or imaginary scenarios. For instance, students might be given a production management dilemma that compares anticipated levels of quality to the meeting of production schedules. Gorman et al. (2000, p. 463) cite the usefulness of introducing "simple, short cases based on minor ethical dilemmas faced by engineers on a daily basis."

Case studies bring to life typical challenges in the realm of design, planning, engineering, production, management, and marketing. Often, when case studies are based on real events, names and dates are changed to "protect the innocent" or to separate the nature of the lesson from actual newspaper headlines. Amazingly, students often inquire about whether certain events actually did take place. Case studies add authentic, realistic examples to the curriculum and much can be learned through the difficult experiences of others.

Another way to challenge students is to draw upon experiences within the local manufacturing classroom or laboratory. Each instructor has, no doubt, witnessed questionable practices (shoddy craftsmanship, suspect quality, waste, abuse, and so on). While using the familiar role-playing approach typical of enterprise and mass-production courses, the treatment of teammates could help bring personnel and leadership issues to the forefront.

Drawing on experiences in the current class allows all students to put the issue into perspective. No one would need to relate the context or scenario as everyone is "on the same page." The instructor can also provide immediate feedback as issues arise. For instance, a questionable management tactic or failed assignment can be ($a$) clearly identified, ($b$) analyzed, and ($c$) feedback suggested for developing a more appropriate course of action. A note of caution is to alter the identities to protect individuals from embarrassment or abuse if incidents from previous classes are cited.

Naturally, the most direct method of addressing ethics is with specifically planned lessons. Time can be dedicated to integrity, honesty, and related practices in each class. Units of instruction might cite corporate attempts to maintain ethical standards, illustrated using either printed materials, videos, or other items produced by companies being studied. Examples of corruption, abuse, and other negative actions can be presented as failed strategies. A guest speaker from a local firm might outline the positive efforts of area manufacturers.

A typical lesson might start with a past scenario, along with a description of some decisions that led to a major disaster. Because industrial accidents and managerial corruption are major news, the headlines and news articles might be available through an online database. Plan to outline the actual events during the class, then cite how and where decisions were made (or not made) that led to the problems. Most students can pick up on obvious errors in judgment or action as hindsight is usually 20-20. The key is challenging students to identify a more appropriate response and encouraging them to employ the same level of insight into their future dealings with people and technology.

### *Goal of Manufacturing Educators*

Modern manufacturers tend to compete with what are called "core competencies," exemplified when a company establishes an admirable environmental policy or has a reputation of outstanding product quality

(Russell and Taylor, 2000). Through a unique trait, the enterprise can gain a competitive advantage over rivals. Although companies once competed solely in the areas of production or marketing, some firms now address ethics as a hallmark of their business.

The same can be said for manufacturing technology programs at the secondary and undergraduate levels. An entire department could implement a strategy to highlight ethical themes and practices. This would certainly position the program quite well in the local school. The core competency could involve a total commitment to people, training, and service through production-based experiences.

Hays, Wheelwright, and Clark (1988) described how this approach works in industry, noting how today's organizations must be prepared to learn new concepts and make proper decisions. Their recommendations were based on the dynamic changes in manufacturing due to global competition. In another instance, Sobek et al. (1998) documented how Toyota implemented a successful mentoring system for engineers and managers. This system allowed a corporate culture to be transmitted in an effective way to propagate a spirit of pride and quality.

Manufacturing is about converting resources into useful products, but numerous issues cloud the daily agendas of productive enterprises. Most challenges involve a person or small group making key decisions along the way. When integrity, reliability, and ethics are concerned, only a select group consistently follow the appropriate pathway. As Zimmerman and Beal (2002, p. 239) noted, "public and private educational institutions should strengthen their programs and extension partnerships to provide technical and managerial assistance to manufacturers." This ensures that the optimal behavior is learned by workers, supervisors, and top management through instruction and mentoring.

## A RETRIEVAL DEVICE

*Description*

The purpose of this activity is to provide students with the opportunity to solve a manufacturing design problem involving adaptive design and embedded ethical and social elements. Students also have opportunities to develop sensitivity to the particular challenges produced by disabilities. This could result in greater awareness and appreciation for accommodations called for in the Americans with Disabilities Act and other similar initiatives aimed at providing equal opportunities for disabled persons.

## Educational Outcomes

1. Students will demonstrate proficiency in manufacturing/engineering design within a design constraints environment.
2. Students will apply and analyze the ethical and social benefits of their adaptive design solution within a manufacturing context.
3. Students will evaluate and apply a variety of manufacturing processes and materials to their design solution.
4. Students will learn to communicate and collaborate effectively within a team environment.

## Materials List

- Modeling and prototyping supplies, including:
  - Dowels
  - File folders
  - Masking tape
  - Rubber bands
  - String
  - Plastic sheet, soft wood pieces, general light hardware, fasteners, and hinges
  - Paper towel rolls
  - Paper cups
  - General laboratory consumable supplies
  - PVC pipe
- Manufacturing supplies, including:
  - Hand tools and machines
  - Work tables
  - Safety protection
  - Materials for jigs and fixtures

*Ethics and the Study of the Designed World*

<u>*Instructions*</u>

From time to time, we all need something that is just out of reach. While playing outside, your ball rolls under the front porch. Inside, the cash you made at your part-time job falls behind the couch, or your ring falls down the sink drain. You drop your car keys and accidentally kick them under the car during a rainstorm. Your grandmother cannot reach the medication she needs. Your little brother cannot retrieve the ingredients he needs to make a healthy snack. At your job, a disabled manager cannot retrieve items from the bottom file drawer.

Each of these items can be retrieved using different methods. However, is there a single device (possibly with attachments) that will help a wide range of people retrieve objects of different sizes and shapes from various places? That is your challenge—to design and demonstrate a device that can accomplish as many of the following retrieving tasks as possible for people with a wide range of abilities.

**Part 1 Activities**

As a corporate design team, your *first major task* is to design and build a prototype device that will enable frail adults, people with disabilities, and even children to accomplish the following operations:

1. Retrieve a nonmagnetic ring from a vertical PVC drainpipe that is 4' long by 3" around (ID).
2. Retrieve a flat, unfolded dollar bill from behind a couch or easy chair.
3. Retrieve a 300 tablet Bayer® Aspirin bottle from the bottom shelf of a closed, overhead kitchen cabinet 73" off the ground; place it on a 36" high table; and then replace it.
4. Retrieve an 18 oz. Box of Quaker Oats® from the bottom shelf (6" off the ground) of a closed kitchen cabinet, place it on a 36" high table, and then replace it.
5. Open the bottom drawer of a file cabinet, retrieve one standard, non-hanging file folder from the middle position among at least 10 file folders; remove it from the drawer; place it on top of a table; and then return it to the same position in the drawer and close the drawer.

Your device must be able to be used for each of the preceding tasks by:

1. An elderly, moderate height (5'4" maximum) frail adult. This person has a weak grip and cannot pick up anything heavier than 5 lbs. He or

she cannot bend at the waist or kneel, but does have use of both hands and arms. Tasks demonstrated for this type of person must be done from a standing position.
2. A person who is in a wheelchair. This person cannot bend at the waist and has the use of only one hand and one arm. He or she can lift 10 lbs.

The device must, at least, be:
1. Safe for the users specified
2. Able to perform the specified tasks by the specified people
3. Portable by the user

### Part 2 Activities

Even a product that accomplishes all of the specified tasks will fail to benefit your company and society if it does not have a market or sufficient investment dollars behind it. Your team's *second major task* is to make investors understand and appreciate your design and to convince them to invest in your design and product so that you can begin full-scale production. For this design challenge, assume that you are addressing a team of engineers, marketers, accountants, production supervisors, and salespeople. Your team's presentation must address the engineering aspects of your design, *and* it must convince potential investors that your product is a viable one in which they should invest.

### Testing

To successfully complete this activity, it is important for each team to research the needs of physically impaired people. Participants will then design, test, manufacture, and demonstrate the capabilities of the products to investors. This includes showing them its flexibility and expandability as well as communicating the social good the product can contribute to the lives of the physically impaired. To do this, teams may want to add tasks that go beyond item retrieval and replacement. They might demonstrate retrieval and replacement of additional items, add special or unique features to the product, and demonstrate how these special features enhance its usefulness.

For purposes of the demonstration, obtaining a couch or file cabinet is not necessary. Teams can simulate a couch with a cardboard mock-up and the file cabinet with a cardboard sliding drawer file box. If cabinets are not available in the classroom, these can be simulated also. Remember, you cannot see through a couch or file cabinet, and you cannot see into a wooden cabinet until the door is open. If a wheelchair is not available, teams can simulate it using a standard classroom chair.

After initial positioning, the device must be set up and demonstrated by the user without additional help. The user must connect or add any extensions, additions, or attachments without assistance from anyone else.

All items, including the device, all props, and equipment, must fit through a standard door (36" × 84").

Team members can simulate a frail adult by stooping or kneeling. Be certain not to exceed the 5'4" height limit. Each presentation should include evidence that these requirements have been met. A taller team member can squat a little if needed to meet the height requirement. The parameters in Part 1 of the instructions are an important aspect of the problem.

*Reflection Questions*

1. What were the ethical and social issues your team most considered during the design phase of your project?
2. How important is it to have a person with a disability consult with your team as you work through the design and manufacturing of your adaptive device?
3. What responsibilities do you think manufacturers of goods have in considering the ethical and social use of their products by persons with disabilities?

## CONSTRUCTION TECHNOLOGIES

*Jack W. Wescott*

The learning activity presented in this section was designed to provide students with an appreciation for the ethical issues associated with construction technology. The materials build on the construction technologies portion of Chapter 3, Ethics and the Design and Development of Technological Systems. Simply stated, this section addresses how the study of ethics can be integrated into construction technology. Although this real-life activity is intended to be incorporated into existing construction activities, such as the fabrication of residential-type structures such as sheds, solar greenhouses, play houses, dog houses, pool houses, and so on, it is also appropriate for other nontraditional structures. Admittedly, the fabrication of residential structures is not new and different to the construction technology curriculum, but the integration of specific content that addresses the ethical issues in construction is significant. The strategy of incorporating ethics into traditional instructional activities is also important because instructors typically have little time to add additional lessons to their already crowded agenda.

The problem is introduced to the students using a fictitious case. If appropriate, instructors have the freedom to adapt the case study to real individuals and specific types of structures. As a group, students are challenged to develop a construction specification package that is typical of most real-world construction projects. The activity would be most meaningful if it could be incorporated into an existing laboratory experience rather than within a theoretical classroom setting. Most importantly, as a result of learning how to prepare the package, the students are confronted with a variety of potential ethical issues and will assess the impact such issues have on the client, constructors, and the community.

## THE ETHICS ARE IN THE SPECS

*Description*

Students participate in development of a set of construction specifications that address several ethical issues. They are then provided opportunities to evaluate specifications prepared by other teams and given opportunities to discuss the potential outcomes of the project.

*Educational Outcomes*

1. Students will develop a construction specification package for the construction of a structure.
2. Students will describe what constitutes an ethical issue.
3. Students will develop a specification package and will identify and explain the ethical issues related to the project.
4. Students will develop written and verbal communication skills.
5. Students will work cooperatively in small groups to solve problems.

## Materials List

- Copy of client narrative and discussion questions for distribution to students
- Reference materials relevant to construction specifications
- Peer evaluation form (student developed) for critique of specifications

## Instructions

1. Divide the class into teams to complete the activities. The number of students on each team can be determined by the instructor.
2. Distribute copies of the client narrative and discussion questions to students and provide opportunities for them to talk about the materials among themselves. Each group should establish strategies for gathering additional information needed and developing the required construction specifications.
3. Each group should prepare a set of construction specifications for the structure. Construction specifications should include, at a minimum, the following:
   a. General information
   b. Name of client
   c. Address of client
   d. Intended use or purpose of the structure
   e. Site plan
   f. Location of the structure on the site in relation to other structures or landmarks
   g. Orientation of the structure on the site
   h. Specification fabrication processes
   i. Foundation
   j. Flooring
   k. Walls
   l. Roof
   m. Selection and purchase of materials
   n. Complete itemized bill of materials

o. Contract sum and payment (if applicable)

p. Site cleanup

q. Disposal of surplus materials

r. Policy site cleanup and disposal of materials

s. Project planning

t. Projected completion date

u. Project timetable

v. Change orders

w. Legal issues

x. Quality of work and local codes

y. Protection of persons and property

4. Each group should prepare a critique form to be used when evaluating a set of construction specifications. The form should include a check to see that legal and ethical issues have been properly addressed.

5. After groups have completed the construction specifications, each group will be provided an opportunity to review and critique a set of specifications prepared by another group. Each group should use the critique form they have developed for this purpose.

6. A concluding class discussion should be used to review the results of this activity. Students can be encouraged to reflect on what kinds of issues were difficult to resolve and how their own work compared to that of the group whose specifications they evaluated.

## *Client Narrative and Discussion Questions*

A local resident of the community has approached your technology education instructor to determine if the class could build a storage structure on his property. Your teacher felt that this would be a meaningful opportunity because the experience would allow students to apply what they were learning in class to a real-life construction problem. The prospective client has agreed to finance the cost of the structure but has requested an itemized cost estimate before committing to the project. At face value, this sounds like a rather simple request. However, in addition to the cost estimate, the prospective client has additional probing questions regarding the project, which are addressed in the following discussion questions.

Does the local building code regulate how close the structure can be placed to the neighbor's garage? The building site, in the rear of the client's property, is quite small, so it is important to place the structure as close to the property line as possible. The client has implied that the neighbor who owns the garage has given him verbal permission to place the shed directly on the property line. The only other possible location would be near a drainage ditch that periodically overflows during heavy rains.

Will the structure be built using materials and processes that are in compliance with the local building code? There would be significant cost savings when purchasing materials if the local building codes were slightly altered. The client feels that it is a small building and will not be inhabited by humans so minor changes in the code are acceptable.

Will the owner be required to obtain a building permit? Likewise, will their property taxes increase? A building permit for the structure would cost approximately $25. The client does not want to file for a permit because the permit would be sent to the tax office and the family's property taxes would increase.

Is it possible to use recycled and environmentally friendly building materials? Although recycled and environmentally friendly materials are available, they are more expensive than the traditional building products. The school has access to some of the old, pressure-treated dimensioned stock that could be used for the foundation of the structure. Although the old, pressure-treated materials are still acceptable by the code, recent television news programs have broadcast special programs describing the hazards of such materials.

Who is liable for the students when working on the project in case of property damage? Who is liable if a student is injured while working on the project? The client is reluctant to increase liability insurance coverage and wants the school to incur all responsibility should an injury occur.

Who owns the surplus materials not used in the construction process? Every construction process has surplus materials left over at the completion of the project. The instructor has indicated that the surplus materials could be used by the school to support other, smaller projects. Likewise, the client would like to use the materials for other projects at the location.

What if the students do not complete the structure? Do the students have an ethical responsibility to complete the project even if the structure is not finished by the time school is dismissed for the summer?

## Instructor Notes

When guiding the students through the case study, there are two kinds of questions that students should consider as they work in their groups. One set of questions probes the literal meaning of the story, the integrity of the narrative. What are the facts? Was anything missing? Was the situation believable, plausible, and coherent? The second set of questions should focus on the rhetorical qualities of the case study. Important to such reflection is the following question: What feelings did the narrative materials evoke in you? It is also important to emphasize the close ties between the facts and feelings of each scenario.

To assist the students as they address the issues of the specifications packet, it is suggested that strategies such as guest speakers and field trips be included to supplement the traditional lectures and group discussions. Potentially informative guest speakers include (*a*) the client, (*b*) building inspector, (*c*) construction lawyer, (*d*) local contractor, and (*e*) real estate agent or broker.

Ethical considerations should be a fundamental part of the decision-making process in construction technology. The learning activity presented in this section was designed to expose students to ethical issues related to the process of building a structure. Finally, it is hoped that the activity provides students with the opportunity to apply and assess their own personal ethical values within the context of a technological problem.

# REFERENCES

Baber, L. (2002). *Consumer reports tests hybrid cars.* Retrieved May 9, 2003, from http://www.wfmynews2.com/2wk/2wk.asp?ID=855

Bellis, M. (n.d.). *The history of the automobile.* Retrieved May 18, 2003, from http://www.inventors.about.com/library/weekly/aacarssteama.htm

Freeman, S. (2003, March 13). Hybrid cars attract more buyers. *Wall Street Journal.* New York, Eastern edition, p. D.3.

Gorman, M., Hertz, M., Louis, G., Magpili, L., et al. (2000, October). Integrating ethics and engineering: A graduate option in systems engineering, ethics, and technology studies. *Journal of Engineering Education, 89*(4), 461-469.

Hays, R. H., Wheelwright, S. C., & Clark, K. B. (1988). *Dynamic manufacturing.* New York: The Free Press.

International Technology Education Association (ITEA). (2000). *Standards for technological literacy: Content for the study of technology.* Reston, VA: Author.

Mulgan, G. (1997). *Connexity: How to live in a connected world.* London: Chatto & Windus Limited.

National Council for the Social Studies. (1997). *National standards for social studies teachers.* Silver Spring, MD: Author. Retrieved February 2, 2003, from http://www.socialstudies.org/standards/teachers/

Parker, J. (2003, February 27). Despite some skepticism, hybrids may lead to buying shift in auto industry. *Knight Ridder Tribune Business News.* Washington, DC.

Poe, J. B. (1983). *The American business enterprise.* Homewood, IL: Richard D. Irwin.

Pond, R. J. (1993). *Introduction to engineering technology* (2nd ed.). New York: Merrill.

Rose, L. C., & Dugger, W. E., Jr. (2002). ITEA/Gallup poll reveals what Americans think about technology. *The Technology Teacher, 61*(6), insert 1-8.

Russell, R. S., & Taylor, B. W., III. (2000). *Operations management* (3rd ed.). Upper Saddle River, NJ: Prentice-Hall.

Sobek, D. K., II, Liker, J. K., & Ward, A. C. (1998, July-August). Another look at how Toyota integrates product development. *Harvard Business Review, 76*(4), 36-49.

Wescott, J. W., & Henak, R. M. (Eds.). (1994). *Construction in technology education.* Council on Technology Teacher Education 43rd annual yearbook. Peoria, IL: Glencoe/McGraw Hill.

Williams, R., & Tollet, J. (2000). *The non-designer's Web book* (2nd ed.). Berkeley, CA: Peachpit Press.

Zimmerman, F., & Beal, D. (2002). *Manufacturing works: The vital link between production and prosperity.* Chicago, IL: Dearborn Trade.

# Ethics in a Global Economic System

## Chapter 8

Archie B. Carroll
The University of Georgia
Athens, GA

Technological literacy is an important goal and is a topic that cuts through and permeates organizational life. The ethics dimension is one of the most important facets of technology, but in some instances it is a dimension that has been overlooked. Ethical sensitivity in terms of who will be helped and who will be hurt is a constant theme that must be revisited if an ethical environment is to be created and sustained. Technologies have benefited people in many ways. There have also been serious unanticipated side effects of technology—problems or effects not anticipated before technologies were implemented. Technologies have produced environmental pollution, depleted natural resources, created technological unemployment, and have created unsatisfying jobs. In effect, technology indeed raises ethical issues for society, for professionals, and for organizations.

In this chapter, we provide an overview of ethics and ethical issues, only one of which is technology, as we consider the various "levels" at which ethics might be thought about and evaluated. We begin at the workforce level of organizations, where most of us begin to appreciate the need for ethics, move to the level of management, then to the corporate level, and finally to ethics in government and global affairs. The purpose is to highlight how this vital topic—ethics—touches people, groups, processes, and decisions.

## WORKFORCE ETHICS

Whether the topic is referred to as business ethics, management ethics, or the more inclusive term, workforce ethics, it is hard to overstate the importance of ethical functioning and practice in organizations of the twenty-first century. For years, some have wondered whether ethics is really all that important in the world of organizations, or stated more bluntly, they have asked, "Does ethics pay?" Considerable documentation could be set forth showing how ethics pays. Predicated upon the idea that companies that "do good" will "do well," it is noteworthy that $1.5 trillion is now invested worldwide according to social or ethical criteria in the so-called ethical or socially conscious investing movement.

But, it is easier in the scandal and fraud environment of the early 2000s to make the point that "bad ethics hurts." If you look at companies such as Enron, WorldCom, Tyco, and Arthur Andersen, you cannot help but see the crushing blow to the organization's profitability, reputation, and survival that is caused by legal and ethical lapses and outright fraud and deception.

Prior to 2001, who could ever have imagined that Enron, a company that had built a fairly decent reputation as an exemplary corporate citizen, would now be "on the ropes" fighting for survival in the face of massive fraud? Enron, a technology-based energy company, and its executives were generous supporters of the community in Houston, their hometown. The company captured international attention by building a power plant in India without resorting to the typical bribery that is expected in so many parts of the world. The company lobbied the Bush administration in favor of an international agreement to address global warming (Vogel, 2002). In spite of all this, the company is now prime material for case studies on excessive greed, fraud, and bad corporate ethics. Enron, along with many other major companies, demonstrated firsthand how organizations can be philanthropic but at the same time grossly unethical (Swartz and Watkins, 2003).

A second case is that of the prominent accounting firm, Arthur Andersen. Once regarded as one of the capstone accounting firms in the world, it now lies in ruins due to ambition, greed, fraud, and deception, having gone out of business because it was no longer trusted. Andersen's conviction of obstruction of justice in connection with the Enron calamity brought an end to the 88-year-old accounting firm (Toffler, 2003).

In spite of the recent corporate scandals, companies have been spending valuable resources on ethics training at an unprecedented rate. Typical of such ethics training is that provided by the Josephson Institute of Ethics with their "Ethics in the Workplace" training seminars. Targeted toward managers, human resource executives, ethics officers, and compliance directors, these seminars train participants in how to meet federal standards for compliance, develop prevention mechanisms called for by guidelines, reduce risks of litigation associated with ethical wrongdoing, establish standards and systems, and create strong ethical cultures or workplace environments (www.josephsoninstitute.org/etw.htm).

Companies today are increasingly looking for ethical employees to hire. According to *Job Outlook 2001*, a survey released by the National Association of Colleges and Employers (NACE), the "ideal" candidate knows how to communicate, is honest and has integrity, possesses teamwork and interpersonal skills, and demonstrates motivation and initiative. For the first time since the survey has been taken, honesty and integrity showed up in the rankings and actually ranked second, just behind communication skills. This moral ingredient is finally taking its rightful place among important job candidate qualities.

For too long, employers have been fixated on human and technical skills as key ingredients of successful employees. Now, they are finally recognizing the importance of honesty, integrity, and ethics among the personal qualities valued for successful employment. Whether strong ethics has a direct payoff or an indirect payoff, companies and managers are coming to realize that high ethics is the course to take and that this idea should not be devalued by requiring it to be directly linked to financial payoffs. Though ethics and economics are often in tension with one another, the ethical high road is clearly in the organization's and society's best interests.

## APPROACHES TO MANAGEMENT ETHICS

Whether dealing with technology issues that have ethical implications or organizational or human resource issues, managers who have the primary responsibility for leadership have been seen to display a variety of different ethical approaches or styles in their decision making and practices.

It can often be difficult to discern whether managers are being ethical or unethical, moral or immoral. In our discussion here, we are equating the terminology of ethics with that of morality, though there might be subtle differences that philosophers or theorists might want to make. In thinking about management behavior, actions, or decisions, it is often impossible to clearly categorize these actions as moral or immoral. In a quest to understand management behavior, a third category is usefully added, that of amorality. There are at least three models of management morality that help us to better understand the kinds of behavior that may be manifested by managers. These three models, or archetypes—immoral management, moral management, and amoral management—serve as useful base points for discussion and comparison (Carroll, 1987, 1991, 2000).

## Ethics in a Global Economic System

The media has focused so much on immoral or unethical management behavior that it is easy to forget or not think about the possibility of other ethical types. For example, scant attention has been given to the distinction that may be made between those activities that are immoral and those that are amoral; similarly, little attention has been given to contrasting these two forms of behavior with ethical or moral management.

A major goal in considering the three management models of morality is to develop a clearer understanding of the full gamut of management approaches in which ethics or morality is a major dimension. Further, it is helpful to see through description and example the range of ethical behavior that management may intentionally, or unintentionally, display. Let us consider the two extreme positions first.

### *Immoral Management*

Let us start with immoral management as this approach is perhaps most easily understood and illustrated. Immoral management is a style that not only is devoid of ethical principles or precepts, but also implies a positive and active opposition to what is ethical. Immoral management is discordant with ethical principles. This view holds that management's motives are selfish and that it cares only about its own or its organization's gains. If management activity is actively opposed to what is regarded as ethical, this implies that management can distinguish right from wrong, and yet chooses to do wrong.

According to this model, management's goals are purely selfish (if the individual is acting on his or her own behalf) or focused only on profitability and organizational success (if the individual is acting as an agent of his or her employer). Immoral management regards the law or legal standards as impediments it must overcome to accomplish what it wants. The operating strategy of immoral management is to exploit opportunities for organizational or personal gain. An active opposition to what is moral suggests that managers would cut corners anywhere and everywhere it appeared useful to them. The key operating question of immoral management would likely be: "Can I gain from this decision or action, or can we make money with this decision or action, regardless of what it takes?"

**Examples of Immoral Management**

Examples of immoral management are easy to identify as they frequently involve illegal actions or fraud. The Frigitemp Corporation, a manufacturer of refrigerated mortuary boxes, provides an example of immoral management at the highest levels of the corporate hierarchy. In litigation, criminal trials, and federal investigations, corporate officials, including the president and chairman, admitted to having made millions of dollars in payoffs to get business. They admitted taking kickbacks from suppliers, embezzling corporate funds, exaggerating earnings, and providing prostitutes to customers. One corporate official said that greed was their undoing. Records indicate that Frigitemp's executives permitted a corporate culture of chicanery to flourish. The company eventually went bankrupt because of management's misconduct.

Another example of immoral management was provided by a small group of executives at the Honda Motor Company. Federal prosecutors unraveled a long-running fraud in which a group of Honda executives had pocketed in excess of $10 million in bribes and kickbacks paid to them by car dealers. In exchange, the executives gave dealers permission to open lucrative dealerships, and they also received scarce Honda automobiles that were difficult for dealers to obtain at the time. Eight executives pleaded guilty and many others were indicted.

In the technological sphere, immoral management might be illustrated by the supervisor who wrongly blamed an error the employee made on a technological glitch. Alternatively, he might have used office equipment to shop on the Internet for personal reasons. Other examples might include copying the company's software for home use, using office equipment to network or search for another job, or accessing private computer files without permission.

## *Moral Management*

At the opposite extreme from immoral management is moral management. Moral management conforms to high standards of ethical behavior and professional standards of conduct. Moral management strives to be ethical in terms of its focus on, and preoccupation with, ethical norms and professional standards of conduct, motives, goals, orientation toward the law, and general operating strategy. In contrast to the selfish motives of immoral management, moral management aspires to succeed but only within the confines of sound ethical precepts—that is, standards predicated on such norms as fairness, justice, and due process. Moral management does not pursue profits at the expense of the law and sound ethics. Indeed, the focus is not only on the letter of the law, but also on the spirit of the law. The law is viewed as a minimal standard of ethical behavior because moral management strives to operate at a level well above what the law requires.

Moral management requires ethical leadership. It is an approach that strives to seek out the right thing to do. Moral management would embrace what Lynn Sharp Paine (1994) has called an "integrity strategy." An integrity strategy is characterized by a conception of ethics as the driving force of an organization. Ethical values shape management's search for opportunities, the design of organizational systems, and the decision-making process. Ethical values in the integrity strategy provide a common frame of reference and serve to unify different functions, lines of business, and employee groups. Organizational ethics, in this view, help to define what an organization is and what it stands for.

**Examples of Moral Management**

A couple of examples of moral management are illustrative. When McCullough Corporation, maker of chain saws—generally regarded as a dangerous product—withdrew in protest from the national Chain Saw Manufacturer's Association because the association fought mandatory safety standards for the dangerous saws, this illustrated moral management. McCullough knew its industry's products were technically excellent, but were dangerous, and had put chain brakes on its saws years before, even though it was not required to do so by law. Later, it withdrew from the Association because this group fought government regulations to make their products safer.

Another well-known case of moral management occurred when Merck and Company, the pharmaceutical firm, invested millions of dollars to develop a technologically advanced treatment for river blindness, a third-world disease affecting almost 18 million people. Seeing that no government or aid organization was agreeing to buy the drug, Merck pledged to supply the drug free forever. Merck's recognition that no effective mechanism existed to distribute the drug led to its decision to go far beyond industry practice and to organize and fund a committee to oversee the distribution.

In the realm of computer ethics, an act of moral management was manifested by The Computer Ethics Institute as it set forth what it calls its "ten commandments of computer ethics" (Goldsborough, 2000). The Institute was proposing a higher standard of ethical behavior and guidelines than those frequently practiced by organizations.

## Amoral Management

There are two kinds of amoral managers: unintentional and intentional. Unintentional amoral managers are neither immoral nor moral but are not sensitive to, or aware of, the fact that their everyday business decisions may have deleterious effects on other stakeholders (Carroll, 1991). Unintentional amoral managers lack ethical perception or awareness. That is, they go through their organizational lives not thinking that their actions have an ethical facet or dimension. Or, they may just be careless or insensitive to the implications of their actions on stakeholders. These managers may be well-intentioned, but they do not see that their business decisions and actions may be hurting those with whom they transact business or interact. Typically, their orientation is toward the letter of the law as their ethical guide.

Intentional amoral managers simply believe that ethical considerations are for our private lives, not for business. These are people who reject the idea that business and ethics should mix. These managers believe that business activity resides outside the sphere to which moral judgments apply. Though most amoral managers today are unintentional, there may still be a few who simply do not see a role for ethics in business or management decision making (Carroll, 1987). Fortunately, intentional amoral managers are a vanishing breed.

### Examples of Amoral Management

An early example of amoral decision making occurred when police departments stipulated that applicants must be 5' 10" and weigh 180 pounds to qualify for being a police officer. These departments just did not think about the unintentional, adverse impact their policy would have on women and some ethnic groups who, on average, do not attain that height and weight. This same kind of thinking spilled over into the business context when firms routinely required high school diplomas as screening devices for many jobs. It later became apparent that minority groups were adversely impacted by this policy and, therefore, outcomes were unintentionally unfair to many persons who otherwise would have qualified for employment.

The liquor, beer, and cigarette industries provide other examples of amorality. Though it is legal to sell their products, they did not anticipate that their products would create serious moral issues: (*a*) alcoholism, (*b*) drunk-driving deaths, (*c*) lung cancer, (*d*) deteriorating health, and (*e*) offensive secondary smoke. A specific corporate example of amorality occurred when McDonald's initially decided to use polystyrene containers for food packaging. Management's decision did not adequately consider the adverse environmental impact that would be caused. McDonald's surely did not intentionally create a solid waste disposal problem, but one major consequence of its decision was just that. To its credit, the company responded to complaints by replacing the polystyrene packaging with paper products. By taking this action, McDonald's illustrated how a company could transition from the amoral to the moral category.

In the arena of technology, it could be argued that the video-game industry has been unintentionally amoral because it has developed games that glorify extreme violence, sexism, and aggression without paying much attention to how these games impact the young people who become addicted to them.

## *Patterns of Management*

There are two possible hypotheses regarding the three models of management morality that are useful for ethics in management.

One hypothesis concerns the distribution of the three types over the management population, generally. This *population hypothesis* suggests that, in the management population as a whole, the three types would be normally distributed with immoral management and moral management occupying the two tails of the curve and amoral management occupying the large, middle part of the normal curve. According to this view, there are a few immoral and moral managers, given the definitions stated previously, but that the vast majority of managers are amoral. That is, these managers are well-intentioned, but simply do not think in ethical terms in their daily decision making.

A second hypothesis might be called the *individual hypothesis*. According to this view, each of the three models of management morality may operate at various times and under various circumstances within each manager. That is, the average manager may be amoral most of the time but may slip into a moral or immoral mode on occasion, based on a variety of impinging factors.

Neither of the preceding two hypotheses have been empirically tested. However, they provide food for thought for managers striving to avoid the immoral and amoral types.

It could well be argued that the more serious social problem in organizations today is the prevalence of amoral, rather than immoral, managers. Immoral management is headline grabbing, but the more pervasive and insidious problem may well be that managers have simply not integrated ethical thinking into their everyday decision making, thus making them amoral managers. These amoral managers are basically good people, but they essentially see the competitive business world as ethically neutral. Until this group of managers moves toward the moral management ethic, we will continue to see businesses and other organizations criticized as they have been in the past several decades (Carroll, 1991).

## ETHICS PRINCIPLES, ETHICS TESTS, AND ETHICAL DECISION MAKING

As managers think about responding to the ethical challenges they face, there are a number of different ethical theories or principles they may choose from to help themselves make better and more ethical decisions. Several major ethics principles include the principle of rights, justice, and utilitarianism. These are often referred to as the "big three" ethics principles. The *principle of rights* seeks to protect stakeholders' legal and moral rights when managers are making decisions. The rights principle expresses morality from the point of view of the individual or groups of individuals who claim to hold various rights. An individual, for example, may claim a right to privacy though no such right has been guaranteed by law.

The *principle of justice* strives to "be fair" to stakeholders; that is, to provide them with justice. It is often called the "fairness" principle. The principle of justice is not always easy to administer, however, because there are a number of different bases upon which justice, or fairness, may be interpreted. For example, *distributive justice* seeks to fairly distribute benefits and burdens. *Compensatory justice* seeks to compensate someone for a past injustice. An example is affirmative action programs that are rationalized on the compensatory justice principle. Finally, *procedural justice* seeks to employ fair decision-making procedures, practices, and agreements. Procedural justice is aimed at ensuring stakeholders, such as employees or customers, receive "due process."

The *principle of utililitarianism* is a consequential principle. That is, it considers the consequences or outcomes of a decision as a measure of determining right action. It reasons that if the outcomes of a decision are good, the action or decision is good. In its simplest form, utilitarianism often strives to produce the "greatest good for the greatest number." Utilitarianism is attractive because it forces us to think about the general welfare. It proposes a standard outside of self-interest by which to judge the value of a course of action. It forces the manager to think in stakeholder terms: What would produce the greatest good in our decision, considering stakeholders such as owners, employees, consumers, the community, and others? A weakness of utilitarianism is that it may ignore actions that may be inherently wrong. It could lead to an "ends justifies the means" type of rationalization.

Other ethics approaches or principles include the principles of caring, virtue ethics, and the Golden Rule. The principle of caring, frequently referred to as feminist theory, holds that traditional ethics principles, such as rights, justice, and utilitarianism, focus too much on the individual self and cognitive thought processes. Caring theory is a perspective in which the person is seen as essentially "relational," not individualistic. Thus, the *relationship's* moral worth becomes more important. That is, the extent to which individuals care for each other through reciprocal concern is the key component in ethical behavior.

Virtue ethics, by contrast to the action-oriented principles we have been discussing, focuses more on the individual decision maker possessing virtues such as honesty, fairness, truthfulness, benevolence, and so on. Virtue theory is a system of thought that is centered in the heart of the person, in our case, the manager. It emphasizes *being* rather than doing. Whereas traditional ethics principles ask "What should I do?" in virtue theory, the focus is on "What kind of person should I become?"

The Golden Rule has long been lauded as one of the most important ethical principles for managers. It essentially argues that you should "do unto others as you would have them do unto you." This principle argues that if you want to be treated fairly, treat others fairly; if you want your privacy protected, protect the privacy of others. The key is impartiality. According to the Golden Rule, we are not to make an exception of ourselves. One reason the Golden Rule is so popular is that it is rooted in history and religious tradition and is among the oldest of the principles of living. Further, it is universal in the sense that it requires no specific religious belief or faith.

In addition to ethics principles as decision-making aids, there are a number of "ethical tests" that managers may use to help them make correct decisions (Carroll and Buchholtz, 2003). An "ethical test" is a short, practical question that a manager can use to help him or her make decisions. No single question is recommended as a universal guide for determining "What should I do in this situation?" However, one or more of the tests may be useful to different individuals in different situations.

*Test of Common Sense*

Here, the decision maker simply asks "Does the action I am getting ready to take really make sense?" If it is not consistent with common sense, do not do it. *Does it make sense for me to steal someone's identity? Could I possibly get away with it?*

*Test of One's Best Self*

Each of us has a self-concept. Most of us could construct a scenario of ourselves "at our best." This test simply requires the decision maker to ask "Is the action or decision I am getting ready to take compatible with my concept of myself at my best?" If it is not, do not do it. *Does it represent my concept of myself at my best when I spy on one of my employees via the computer?*

*Test of Making Something Public*

If you are about to engage in a questionable practice or action, you might pose these questions: "How would I feel if others knew I was doing this?" "How would I feel if all my family, friends, and colleagues knew I was doing this?" "How would I feel if this were covered on the evening news tonight for the entire world to see?" If you are uncomfortable, do not do it. *How would my family and friends perceive me if they found out I copied the company's software for personal home use?*

*Test of Ventilation*

The idea here is to expose your proposed action to others and get their thoughts on it before you do it. This test works best if you get opinions from people you know might not see things your way. "Ventilate" the issue;

get others' points of view to help you clarify your own thinking. *John, do you think it is okay for me to monitor Jane's e-mail without her knowing about it?*

## Test of the Purified Idea

It is easy to think a course of action is "purified" because your boss or someone in authority, such as an accountant or lawyer, says it is okay. The key point to remember here is that this does not necessarily make the course of action right. It may still be wrong or questionable and you, not they, will be held responsible for it. *Well, Brad, the head of the computing office, said it was okay if I accessed some private files without permission, just as long as I didn't tell him!*

These ethical tests do not embody moral thinking in and of themselves. However, when used in the right circumstances, the manager may find one or more of them to be useful in keeping on the correct path.

The previous ethics principles and ethics tests may be of great assistance to managers seeking to improve their ethical decision making. At this juncture, however, let us discuss ethical decision making as a process. Decision making is at the heart of the management process. If there is any practice or process that is synonymous with management, it is decision making. Decision making usually involves a process of stating the problem or decision to be made, analyzing the problem or decision, identifying possible courses of action, weighing the pros and cons of the courses of action, deciding on the best alternative, and then implementing the decision.

Decision making is a challenge to managers, and decisions that contain ethical overtones or implications are especially difficult. After we leave the relatively value-free decisions, such as which piece of equipment to buy or which software to install, decisions quickly become complex, especially when they have a value component.

The goal in ethical decision making is to adopt an approach that ensures ethical due process. There is no way to guarantee that a decision-making approach will result in the most appropriate decision considering the mix of business and ethics considerations that inevitably one faces. Ethical due process, however, is achievable if the decision maker strives to be certain that ethical factors and dimensions are considered in the process.

One approach to ethical decision making is to employ the ethics theories, principles, and ethics tests discussed earlier. These principles are designed to help decision makers arrive at the most ethical decision or outcome.

Another approach to ethical decision making is to ask and systematically answer a series of questions designed to surface and deal with ethical considerations. Laura Nash (1981), for example, posed a list of 12 questions that should be employed to assist in ethical decision making:

1. Have you defined the problem accurately?
2. How would you define the problem if you stood on the other side of the fence?
3. How did this situation occur in the first place?
4. To whom and to what do you give your loyalties as a person and as a member of the corporation?
5. What is your intention in making this decision?
6. How does this intention compare with the likely results?
7. Whom could your decision or action injure?
8. Can you engage the affected parties in a discussion of the problem before you make your decision?
9. Are you confident that your position will be as valid over a long period of time as it seems now?
10. Could you pose, without qualms, your decision or action to your boss, your CEO, the board of directors, your family, or society as a whole?
11. What is the symbolic potential of your action if understood? If misunderstood?
12. Under what conditions would you allow exceptions to your stand?

An approach to ethical decision making that is similar to Nash's 12 questions is an approach based upon a series of shorter, more pointed questions. Three questions, framed as an "Ethics Check," were formulated by Blanchard and Peale (1988) in their popular book, *The Power of Ethical Management*. The questions were (*a*) Is it legal? (*b*) Is it balanced? and (*c*) How will it make me feel about myself?

A similar approach, using short questions and guidelines, has been employed by Texas Instruments. They term their questions as an "Ethics Quick Test." The questions include:

1. Is the action legal?
2. Does it comply with our values?
3. If you do it, will you feel bad?
4. How will it look in the newspaper?
5. If you know it's wrong, don't do it.
6. If you're not sure, ask.
7. Keep asking until you get an answer.

Texas Instruments publishes these questions and statements on a small wallet card that their managers and employees may carry with them at all times.

A final set of short questions is that employed by Sears, Roebuck and Company (1997). Sears refers to these as its five "Guidelines for Making Ethical Decisions":

1. Is it legal?
2. Is it within Sears' shared beliefs and policies?
3. Is it right/fair/appropriate?
4. Would I want everyone to know about this?
5. How will I feel about myself?

Each of these sets of questions or statements, many of which are similar, are intended to produce a process of ethical inquiry that is of immediate use and understanding to a group of managers and employees. These questions help ensure that ethical due process will take place in a decision-making situation. They cannot tell us whether our decisions are ethical, but they can help managers be certain that they are raising the appropriate issues and are genuinely attempting to be ethical.

One might ask at this juncture what all of this has to do with technology education and technological literacy. If technology education were about preparing people for technical occupations, this discussion might be out of place. However, because technology education is about general education, and many technology education students will one day hold positions of responsibility in management, students should be exposed to the processes available for guiding ethical decisions.

A strategy for teaching ethics that has appeared numerous times in this book is the use of case studies and scenarios to provide opportunities for classroom discussion about ethics and ethical decisions. The lists of questions presented in this section can be useful for students to consider as they analyze scenarios involving management decisions. Perhaps by imagining themselves in roles of responsibility within the context of these discussions, using the guiding questions to think through what should be done, students will develop the sensitivity and wisdom needed to deal with ethical issues as they face them later in life.

## CORPORATE-LEVEL ETHICS

Just as many actions can be taken at the managerial level to ensure ethical decision making and practice, there are also actions and policies that can be invoked at the corporate level to help ensure ethical behavior. Sometimes, it is difficult to differentiate between the management and corporate level, however, as the two are intertwined. At the corporate level, the issues that corporations face are frequently framed as matters of corporate social responsibility or, more recently, corporate citizenship.

Corporate citizenship, as a concept, may be framed narrowly wherein it just focuses on community involvement or more broadly wherein it is concerned with the whole host of economic, legal, ethical, and philanthropic expectations placed on the firm at both the domestic and global levels. Today, there is an expectation that corporations and all businesses will be good corporate citizens. The concepts of corporate social *responsibility* (CSR), *responsiveness,* and *performance* are part of the umbrella of concerns falling under the rubric of corporate citizenship. Corporate social responsibility emphasizes the *obligation* and *accountability* of firms, corporate social responsiveness emphasizes *actions* and *activity,* and corporate social performance emphasizes *outcomes* or *results.* Taken together, the message society has been sending business firms is that they exist within the society and have a responsibility to the society in the ethical and social spheres as well as in the financial spheres.

Clear evidence exists that the business community is increasingly accepting these responsibilities in the realm of corporate citizenship. For example, in 1992, a new organization was formed called BSR—Business for Social Responsibility. According to BSR, it was formed to fill an urgent need for a national business alliance that fosters socially responsible corporate policies. In 2002, BSR claimed over 1,400 business members, including such notable names as Levi Strauss, Stride Rite, Hasbro, Reebok, Honeywell, Lotus Development Company, Timberland, and hundreds of others. More about BSR may be learned at their Web site located at http://www.bsr.org.

Another example of the extent to which corporations are embracing CSR is seen in the annual "corporate citizenship" awards given out by *Business Ethics* magazine. In spring 2003, for example, *Business Ethics* magazine presented its 100 best corporate citizens rankings. The number one ranked company in 2003 was General Mills. The company's two prominent areas of excellence are service to the community and to women and minorities. Other "corporate-level" recognitions for corporate citizenship include *Fortune's* annual "most admired corporations" rankings and the Council on Economic Priorities' annual "corporate conscience" awards.

In 2002, The Conference Board released a report documenting that corporate citizenship is "becoming a central concern at leading companies." The report indicated that, more and more, companies are accepting corporate citizenship as a new strategic and managerial function that carries with it bottom-line repercussions. The report went on to document that nearly 90 percent of corporate-level managers report that their companies have a citizenship goal as part of a statement of core values or business principles. Although traditional corporate relations, community affairs, and contributions programs tend to dominate, an emphasis on broader citizenship relations, including the environment and sustainable development, is emerging as a new model (Conference Board, 2002).

Highlighting the relationships between the business sector and corporations and the rest of society, the primary reasons for the expanded notion of corporate citizenship includes globalization, the worldwide expansion of business, private enterprise, and the market economy and heightened expectations from society and consumers that business can and should fill needs formerly left to governments and should better align shareholder and stakeholder interests.

At the corporate level, companies are increasingly addressing social concerns and making them a part of their corporate level strategies. Ben & Jerry's made its name by focusing on environmental responsibility, Nike learned the hard way what the price was of perceived injustices at overseas factories, and Shell Oil encountered a public-relations nightmare when

Greenpeace activists embarked on a high-profile protest against the dumping of oil in the North Sea. Consequently, corporate responsibility has gone global, and now more than 300 firms worldwide have signed on to the United Nations Global Compact in which they have pledged good corporate citizenship in the arenas of human rights, labor standards, and environmental protection (Vogel, 2002).

Although increasingly addressing corporation-world-society issues such as these, companies have also taken management actions to shore up their ethics within their companies. In addition to the guidelines for ethical decision making, already discussed, there are a number of *organizational actions* that corporate leaders are taking to ensure that they have ethical climates in their organizations. These usually include policies, mission or values statements, or corporate conduct guidelines. When McNeil Labs, a subsidiary of Johnson & Johnson, voluntarily withdrew Tylenol from the market immediately after the reports of tainted, poisoned containers, some people wondered why they made this decision as they did. An often cited response was, "It's the J & J way." This policy statement conveys a significant message about the firm's ethical climate.

Because the behavior of managers has been identified as the most important influence on the ethical behavior of organization members, it should come as no surprise that most actions and strategies for improving the organization's ethical climate must emanate from top management and other management levels as well. The process by which these kinds of initiatives have taken place is often referred to as "institutionalizing ethics" into the organization. Some of the "best practices" for creating an ethical climate or culture in an organization include the following, with top management leadership serving as the hub of all others (Carroll and Buchholtz, 2003):

- Top management leadership (moral management)
- Effective communication of ethical standards/expectations
- Ethics officers and programs
- Realistic objectives that do not thwart ethical behavior or induce unethical behavior
- Ethical decision-making processes (as discussed earlier)
- Codes of ethics or conduct
- Whistle-blowing mechanisms for those observing questionable activities

- Disciplining of violators of ethics standards
- Training and workshops in business ethics
- Ethics audits and assessments

## ETHICS IN GOVERNMENT

Ethics as a component of technological literacy is not only important at the workforce, management, and corporate levels, but at the government level as well. One major reason for this is the role the national and state governments play in the approval of products and processes involving technology. In addition, the federal government increasingly plays a role as corporations get involved in country level or global business interests. There is also global interest in technology transfer between and among nations. Consequently, it is natural to consider ethics in government as a part of ethics in the global economic system.

To gain an appreciation of the role of ethics in the institution of government, it is useful to reference several organizations that embrace responsibility for ethics in government. There is a Center for Ethics in Government, for example, which is sponsored by the National Conference of State Legislators in Washington, D.C. (www.ncsl.org/). This ethics center claims responsibility for a number of issue areas that are affected by ethics in government. Some of the issue areas include agriculture, banking and finance, health and human services, and information technology. In the information technology category, specific ethical issues are addressed. Among the topics of concern under information technology are the following:

- Computer crime
- Electronic commerce (e-commerce)
- Identity theft
- Information privacy
- Integrated Criminal Justice Information Systems
- Internet and e-mail
- Legislative information technology
- Telecommunications

Another important organization is the U.S. Office of Government Ethics. This small agency, housed in the Executive Branch, was established by the Ethics in Government Act of 1978. Some of its program topics, all of which address ethical issues faced in the governmental sphere, include (*a*) gifts from outside sources, (*b*) gifts between employees, (*c*) conflicting financial interests, (*d*) financial disclosure, (*e*) misuse of positions, (*f*) outside activities, and (*g*) conflicts of interest. At the state level of government, practically every state of the union has some form of ethics office or commission. For example, there is the Massachusetts State Ethics Commission and the Georgia State Ethics Commission. All of these organizations and agencies, whether at the federal, state, or local level, address ethical dimensions involving technology.

The role of ethics in government becomes particularly acute at the global level. The global level of commercial transactions, for example, is affected by both home country and host country laws, regulations, and conventions. From the standpoint of the United States, for example, the Foreign Corrupt Practices Act governs bribery and corruption for U.S. firms doing business in other countries. For years, this was one of the few ethics legislations that governed international transactions. Today, by contrast, we are observing expanded activity as such organizations as the World Trade Organization, the International Monetary Fund, and the World Bank join forces as international bodies that set the rules for the global economy.

Ethics issues between countries and trading partners has become an enormously important topic as a growing anticorruption movement has begun around the world. One bit of evidence of this anticorruption movement is the emergence of an organization called *Transparency International,* modeled after the human rights group, Amnesty International. Transparency International has established itself as the world's foremost anticorruption lobby. The other major initiative at the global level is the Organization for Economic Cooperation and Development (OECD), with over 30 member countries, agreeing to ban and fight international bribery and asking each member to introduce laws patterned after the U.S. Foreign Corruption Act (Carroll and Buchholtz, 2003).

These are just some of the ethics concerns that arise at the governmental and global levels. With technology constantly on the table as a flash point, issues of international competitiveness, protectionism, industrial

## Ethics in a Global Economic System

policy, political risk analysis, and antiterrorism are other major topics of paramount significance involving international stakeholders.

It is essential for technologically literate citizens in the twenty-first century to be engaged with ethical issues in government. A democracy is shaped by the involvement of its citizens, and ethics should be a factor in the choices citizens make. Technology education has a significant role to play as it educates learners about the technology that drives the world's economic engine and also raises awareness of the impacts of that technology on society and culture.

## SUMMARY

The subject of ethics in a global economic system is a vast topic. In this chapter, we have provided a brief overview of ethics topics and issues beginning with the lowest organizational denominator—workforce ethics. At a next higher level, we have discussed both management and corporate ethics. Finally, the topics of ethics in government and ethics at the international level have been touched upon.

A major component of technological literacy is ethical literacy. It could be argued that the two most important driving forces of change in the world economy over the past decade have been technology and global competition. Consequently, it is expected that the issues of ethics and technology, conceptualized at a global level, will be dominant elements in education for some time to come. It should come as no surprise, therefore, that books such as Paul Alcorn's *Practical Ethics for a Technological World* (2001) has been written, and this type of book will continue to be relevant and timely for technological literacy. For it is at the nexus of technology and human values that the most challenging and important questions facing humankind will be raised in the future.

## INSTRUCTIONAL ACTIVITIES

*Exercise 1*

1. Select a technology that you are familiar with and think about its design and use. This technology could be something you use in the

office, such as computer, voice mail, Internet, cell phone, palm computer, or some other technology.
2. Make a list of the ethical issues that arise in your mind regarding this technology's design and use. *Hint:* Who can be hurt by the technology and how?
3. Identify ways these ethical issues you have addressed could be dealt with.

*Exercise 2*

Following is a list of different technologies that are in existence today. Address the question of whether each is ethical. What are the potential ethical issues embedded in each?

- Automobiles
- Computers
- Cell phones
- Hand guns
- Rat poison
- Television

*Exercise 3*

Following is a list of activities that may be accomplished with technologies commonly found in organizations today. Identify the ethical problems that may be inherent in each of these activities:

- Monitoring
- Recording/taping
- Surveillance
- Photographing
- Measuring
- Watching
- Motivating
- Leading

## Exercise 4

Identify the major ethical issues and implications of the following technologies that are present in the workforce today. How may they be used for harm? How many are used inappropriately? Unethically?

- E-mail
- Internet
- Cell phones
- Software

## Exercise 5

If you were asked to develop "ten commandments for computer use," what would the five most important commandments be? Why?

## Exercise 6

What are the ethical issues surrounding the following topics that generally fall within the definition of bioethics?

- Genetic engineering
- Stem cell research
- Cloning
- Therapeutic cloning
- Genetic testing and profiling
- Genetically modified foods

## Exercise 7

Using the definitions of immoral, moral, and amoral management, as defined in this chapter, research and generate at least two examples in each category that involve technology or technology-based products or processes. Demonstrate how each example fits the criteria of the management category.

# REFERENCES

Alcorn, P. A. (2001). *Practical ethics for a technological world.* Upper Saddle River, NJ: Prentice Hall.

Blanchard, K., & Peale, N. V. (1988). *The power of ethical management* (p. 20). New York: Fawcett Crest.

*Business Ethics.* (2003, Spring). 100 best corporate citizens, 6-10.

Carroll, A. B. (1987, March/April). In search of the moral manager. *Business Horizons,* 7-15.

Carroll, A. B. (1991). The pyramid of corporate social responsibility: Toward the moral management of organizational stakeholders. *Business Horizons, 34,* 39-48.

Carroll, A. B. (2000). Ethical challenges for business in the new millennium: Corporate social responsibility and models of management morality. *Business Ethics Quarterly, 10*(1), 33-42.

Carroll, A. B., & Buchholtz, A. K. (2003). *Business and society: Ethics and stakeholder management* (5th ed.). Cincinnati, OH: South-Western.

Conference Board. (2002, August 15). Corporate citizenship programs gaining more attention among CEOs and top managers (press release).

Goldsborough, R. (2000, January/February). Computers and ethics. *Link-Up, 17*(1), 9.

Nash, L. (1981, November/December). Ethics without the sermon. *Harvard Business Review,* 79-90.

Paine, L. S. (1994, March/April). Managing for organizational integrity. *Harvard Business Review,* 106-117.

Sears, Roebuck & Company. (1997). *Code of business conduct,* 2.

Swartz, M., & Watkins, S. (2003). *Power failure: The inside story of the collapse of Enron.* New York: Doubleday.

Toffler, B. L. (2003). *Final accounting: Ambition, greed, and the fall of Arthur Andersen.* New York: Broadway Books.

Vogel, D. (2002, August 20). Recycling corporate responsibility. *Wall Street Journal,* B2.

# Closing Thoughts about Ethics for Citizenship in a Technological World

Chapter 9

Roger B. Hill
The University of Georgia
Athens, GA

This book grew out of a belief that ethics is an important part of technological literacy. The editor and authors understand that for a person to be a responsible citizen in today's world, ethics must influence decisions and guide actions. The *Standards for Technological Literacy: Content for the Study of Technology* (ITEA, 2000) are intended to guide *all* educators—not just those in the field of technology education—in helping students to become technologically literate. Teachers in any field could benefit from much of the content in this volume, but it is likely that professionals in the field of technology education will be the primary audience.

The authors of this book have provided a variety of perspectives on the topic of ethics, but have also tried to provide practical materials that will help teachers and teacher educators emphasize ethics as a part of the content in their classes. Technology educators have a rich context in which to help students learn to solve ethical dilemmas as they teach students technological problem-solving skills.

In Chapter 1, a decision-making model was presented as a guide for resolving ethical problems that might arise within a technological context. The other chapters in this book have provided an array of materials describing the role of ethics internationally, ways in which ethics influence the designed world, ethical impacts on society, how ethics and character develop, ways technology education already addresses ethics, and the role of ethics within a global economy. As a part of this concluding chapter, we once again turn to the model for making ethical decisions and provide an example of how it might be applied in solving a dilemma faced by a technology teacher.

## THE SCENARIO

It was nearing the end of the spring semester at Midtown High School, and Misty Jones was staying busy as a second-year technology teacher. She had graduated from a nearby university with a Master's degree just two years ago, but had worked as an engineer for Acme International, a major manufacturer of polymers and composites, before deciding she wanted to become a technology teacher.

Misty's undergraduate degree was in chemical engineering, and she had enjoyed her work in that field. What she did not like was moving seven times in her first four years of employment, as the corporation placed her at different plants across the country. Her position involved solving production line problems, and it appeared that her job would continue to require frequent moves. Changing her career path to that of being a technology teacher allowed her to continue using her engineering knowledge, but placed her in a situation in which she could establish some roots within a community.

The courses Misty taught were part of a newly implemented technology education program at Midtown that had an engineering design focus. The curriculum guided students to understand the basic components of engineering design—representing solutions with narrative, graphics, analytical computations, and physical artifacts. Course work included Introduction to Technology, Engineering Graphics, Research and Design, and Engineering Applications. The technology curriculum was also strongly coordinated with coursework in mathematics and science.

Establishing an advisory committee for the new curriculum had led Misty to become involved in the *Partners in Education* program at Midtown High. This initiative matched community businesses with local schools and provided opportunities for them to support various educational activities. A success story Misty was especially proud of was a $5,000 scholarship for graduating seniors who were planning to enroll in a university engineering program. The scholarship was underwritten by Acme International, the corporation for which Misty had once worked. It also provided a very nice goal for her technology education students to work toward.

Misty was responsible for managing the selection process for the Acme scholarship, and this year, there were several outstanding finalists. A committee had been formed to review applicant paperwork, and it included two other teachers, two engineers from the local society of professional engineers, a university faculty member who had been Misty's major professor, and two representatives from Acme International.

As the scholarship selection process came to a close, the final votes from the selection committee resulted in a tie between two candidates. Some committee members had indicated they might abstain from a final vote if they felt they had a conflict of interest. One of these two finalists was the son of a well-known member of the community, so Misty assumed that a committee member had abstained when she saw there was a tie vote. The selection rules called for her to make the final decision in case of a committee tie.

Both of the final candidates for the scholarship were outstanding students. One was a minority student who had struggled some academically, but through hard work and support from several teachers had successfully completed the entrance requirements for a biomedical engineering program. The other candidate had completed entrance requirements for an aerospace engineering program and was also senior class president, Beta club treasurer, and captain of the debate team.

Misty decided to talk with her department chairperson about the scholarship decision. Together, they weighed the pros and cons for each candidate. One would be the first in his family to attend college and would make an outstanding role model for other minority students. The other came from a very successful family, and yet was humble and personable in his dealing with people. He was a leader among his peers, but was very mature for his years. Misty's colleague, however, encouraged her to award the scholarship to the minority student.

After talking about the situation for 45 minutes, Misty told her department chairperson she was going to award the scholarship to the minority candidate. The two of them agreed that this would be a good use of the funds, although both students were very deserving of the award.

The scholarship was to be awarded at the senior honors banquet. Everyone in attendance was pleased when Misty announced the winner and handed him a small plaque, telling him there would be a $5,000 check in the mail. The losing candidate was very gracious, although his disappointment was evident. He congratulated his friend and stood with him as people came by to speak to them at the end of the ceremonies.

When Misty returned home that evening, she sat down to check her e-mail as was her custom. She normally used an account she had with the school system, but this evening she also decided to check an account on Yahoo! that she had used years ago when she was traveling a lot. She

## Closing Thoughts about Ethics for Citizenship in a Technological World

periodically logged on to that account to clean it out, but her colleagues at school sent things to her new account, and she didn't give the Yahoo! e-mail address out to anyone anymore.

As Misty scanned the list of messages in her inbox, one of them caught her attention. It was from one of the scholarship committee members at Acme International. This person had been unexpectedly called on to troubleshoot some problems at one of the company's processing plants in Mexico and had not had time to let Misty know he would be away. He was e-mailing to provide his vote on the scholarship decision, but his laptop computer had Misty's Yahoo! e-mail address in the address book—not the school e-mail address that she was now using. His vote was for the young man who had not been awarded the scholarship, but his vote would have made a difference in the committee decision because the other votes were tied.

Misty stared at the e-mail with a growing sense of concern. She then checked the time and date on the message. It had been sent prior to the end of the time period for committee member votes to be in, but due to differences in time zones, it had reached Misty's Yahoo! account a few minutes past the deadline.

Now, Misty wondered what to do. The next day at school, she was to have a meeting with both of the finalists to provide feedback from the scholarship committee on their applications. This was something that had been written into the competition so that finalists could further benefit from the process. There would be other opportunities for both candidates to apply for awards, jobs, and promotions, so constructive feedback might help them in future situations.

The reality was that if Misty had checked her alternate e-mail account just a day earlier, she knew the scholarship would have been awarded to the other student. She continued to think through the circumstances and realized she would have to decide whether to keep quiet about what had happened or come forward with the information. If she told others what had happened, the scholarship might be taken from the student who had received it and awarded to the other finalist. This would be a devastating blow to the student who had received the award. She had overheard comments after the honors night program that indicated the student would not be able to enroll in college if it were not for this financial assistance.

What should Misty do? How could she meet with the students tomorrow and provide feedback from the scholarship committee without acknowledging the mistake? In addition, if the committee members later compared notes on how they had voted, the circumstance might be discovered. If she kept quiet, would this perhaps jeopardize the future availability of the scholarship? The committee member who e-mailed the vote was, in fact, instrumental in deciding that Acme International would fund the award and also had the power to redirect these funds to some other worthy cause.

## APPLYING THE DECISION-MAKING MODEL

The preceding scenario provides opportunities for a number of the concepts presented throughout this book to be applied. It includes many of the features that make ethical decisions among the most difficult

Figure 9-1. Revised Kidder Model for ethical decision making in technology education (Baker, 1997; Kidder, 1996).

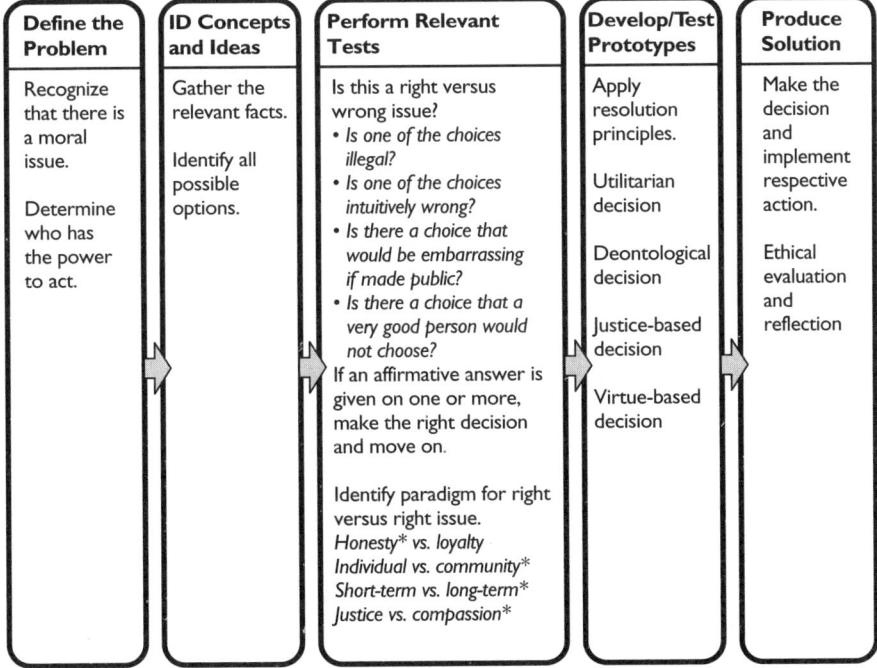

* Kidder identified four dilemma paradigms that are usually a part of right versus right dilemmas. The values identified with asterisks are those Kidder would recommend choosing in situations in which a pair of competing values appear to have equal weight.

## Closing Thoughts about Ethics for Citizenship in a Technological World

choices that we, as human beings, have to make. Although it is not unlike some of the case study activities presented in Chapter 3, it is provided here, along with another copy of the model for ethical decision making from Chapter 1, to provide some closing opportunities for reflection on what this book is really all about—the making of ethical decisions within a complex technological world.

In considering how the decision-making model applies to Misty's situation, it is first important to recognize that there is a moral issue involved in her circumstances. That is why Misty is struggling with what to do. She is a competent engineer, and if this was about choosing the proper chemical to precipitate a particular ingredient in a solution, she could determine a straightforward course of action based on scientific facts. The decision she faces is not so easy. Although it was brought on by the modern marvels of the Internet and electronic mail, the ethical choices presented are not unlike the choices that people have struggled with for generations.

Does Misty have the power to act in this situation? Yes, she does. She can explain what happened and change the recipient of the $5,000 check she will be requesting for the winner. She can also choose to not take any action and pretend that nothing happened, hoping that the scholarship committee members never compare notes and learn that no one abstained from the selection vote.

Misty has most of the relevant facts already at her disposal. She might explore other alternative solutions. One of those would be to make two awards this year, but she does not have the resources to do that. She also does not believe it feasible that she could find an additional $5,000 from some other source.

There are four criteria that can help a person determine whether a decision is a right versus wrong choice. In this particular case, none of the choices Misty is considering are illegal, none are intuitively wrong, and it is unclear what choice a very good person would make. Although the situation has great potential for embarrassment, the source of embarrassment is the action that brought her to this point and not the choice she is about to make.

It appears that the situation Misty finds herself in is a right versus right situation. A decision to avoid changing the award recipient might reflect some level of loyalty to her department chairperson, but honesty calls for her to reveal what has happened. There are individual interests on the part of the students involved, but there may be a community interest involving the minority candidate. Without the scholarship, he will probably not attend college; the other student's parents are financially able to pay for his college expenses. In the short-term, it might be easier to avoid the embarrassment of making a change in the award, but in the long-term, other students might lose the benefit of this scholarship if the mistake is discovered and the scholarship is no longer funded. A justice perspective might call for honoring the democratic process of the committee vote, but compassion for the students might lead Misty to avoid a decision that would be very upsetting to one of the students along with his family and friends.

Resolution principles listed in this model consist of deontology, utilitarianism, and justice and virtue theories. There are some additional perspectives described in Chapters 2, 4, and 5 of this book that might also be considered for use in this stage of the process, but for purposes of this illustration, the four listed in the fourth stage of the model are discussed.

If Misty makes a decision from a utilitarian resolution perspective, she will ask herself which choice will provide the greatest good for the greatest number of people. This line of reasoning might lead her to again consider the positive influence the student who received the scholarship plaque would have on other young people. There are many adolescents, especially within the minority community, who need role models who can inspire them to believe in their own potential. If this student can go to college and succeed in becoming an engineer, it could have a significant impact on a lot of people. If Misty makes an ends-based decision, she might choose to remain silent. In any case, her choice will be one that avoids rescinding the scholarship award she has already made.

From a rules-based perspective, operating from deontological resolution principles, Misty would decide what to do based on a desire to be consistent in upholding high ethical standards. These standards would surely include absolute honesty and integrity. This line of thinking would leave Misty little choice about telling the truth about what had happened, and she would need to try to correct the error that had been made. She would need to tell the students the truth and explain that the award would go to the student who had received the most votes from the scholarship committee. All of this would be done with hopes that things would work out for the other student, but that would not be something she could control.

To make a justice-based decision, Misty will need to consider what social contracts might be a part of the scholarship decision process. The use of the word "justice" here is different from the meaning used in the "justice vs. compassion" statement of Stage 3 in the model. Misty might consider the hardships placed on the minority student because of his skin color and make a choice in his favor to compensate for previous wrongs. Justice-based thinking can lead to actions intended to correct what are perceived to be past injustices.

Misty could choose to make a virtue-based decision. This would involve trying to place herself in the shoes of both students and attempting to take actions that would treat both of them with dignity and respect. She might sit down with the students, explain to them what had happened, and give them an opportunity to help resolve the situation. This would involve risks—they might disagree and become angry. Another possible outcome would be for the student with the most committee votes to concede the prize to the student who had been announced as the winner. Whatever the outcome, a virtue-based decision would demonstrate admirable character traits.

After Misty has made a decision, she will have to act on it. The choice will not be easy, but if she has carefully thought through the situation, applying the steps of the decision-making model, she will be able to present a rational defense of her actions. She will also be able to be more consistent in her approach to ethical decisions as she applies this model to other dilemmas that arise, whether at work or in her personal life.

## SUMMARY

Although the ethical decision-making model can be a helpful tool for thinking through difficult decisions, applying ethics to the problems that come up in life is a challenging activity. No plan is perfect, and although the model can help provide structure, there will be times when mistakes are made. Errors are often identified in retrospect, but it is important to learn from mistakes and move on.

A key outcome that educators should seek as they work with students is for them to have some understanding of ethics, appreciate the significance of ethical decisions in a technological world, and be committed to upholding high ethical standards themselves. Having teachers and adult role models who demonstrate ethical behavior is important for this goal to be achieved. Including an emphasis on ethics as a part of instructional activities is also helpful. To that end, this book has provided an array of materials for use by teachers and teacher educators.

There are many additional books and resources that can enrich one's understanding of ethics and ethical problem solving. Some of these have been identified on the companion Web site for this book located at http://www.uga.edu/teched/ethics. Others can be found in public libraries and school media centers. The materials in this book are by no means comprehensive, but they do provide an overview of ethical philosophies, ethical decision making, and examples of how ethics can be included in instructional activities.

In conclusion, the marvels of modern technology have provided many wonderful tools that can make our lives easier. These marvels can also make our lives miserable. The choices we make about how technology is put to use often involve ethical decisions. As educators, we should make every effort to be certain the students we come into contact with are not only knowledgeable about technology, but are also equipped to use that technology in a responsible and ethical manner. This will only happen as we demonstrate the importance of ethical behavior in both word and deed.

## REFLECTION QUESTIONS

1. In the scenario presented in this chapter, what should Misty do?
2. Which of the resolution principles in Stage 4 of the decision-making model is most consistent with your personal belief?
3. What other philosophical perspectives presented in this book appeal to you and would provide a sound basis for making ethical decisions?
4. What ethical dilemmas have you recently faced, and how might the ethical decision-making model have been applied to these circumstances?
5. If ethics are included as part of the content in a class at school, how should students be tested on this material?

# REFERENCES

Baker, S. (1997). Applying Kidder's ethical decision-making checklist to media ethics. *Journal of Mass Media Ethics, 12*(4), 197-210.

International Technology Education Association (ITEA). (2000). *Standards for technological literacy: Content for the study of technology.* Reston, VA: Author.

Kidder, R. M. (1996). *How good people make tough choices.* New York: Simon & Schuster.

# INDEX

## A

Abel, P., 59
Accommodation ethic, 36
Acts of mercy, 50
Adams, John, 26
Adler, I., 40
Africa
    capacity for sustainable development, 34
    corruption in, 33
    ethics in, 33–35
    legal system and moral code, 34
    religion, 35
Agricultural biotechnology
    ethical issues in, 65–70
Akerele, O., 71
Alamäki, A., 175
Alcorn, P. A., 153
Alternative energy, 133
Americanization, 30
Amin, A. A., 153
Amoral management, 249–250
    examples of, 249
    intentional, 249
    unintentional, 249
Andrew, T., 36
Animal cloning, 69
Annas, G. J., 130
Aristotle's golden mean, 15
Aronson, S. M., 51
Arthur Andersen, 244
Artificial intelligence, 151
Asia
    Buddhism, 35
    caste system, 37
    Confucianism, 35
    ethics in, 35
    Hinduism, 37
    Shintoism, 25
    suppression of women, 37
    *Tao*, 36
    Taoism, 35
    Zen Buddhism, 36
Assimilationist theory, 30
Authority, 2
    managing, 154

## B

Baber, L., 222
Backer, P. R., 42
Baker, A. J. M., 76
Baker, S., 13, 271
Barker, R., 167
Battelle, P., 55
Beach, M. C., 52
Beal, D., 231
Bekes, C. E., 52
Belief systems, 154
    closed societies, 155
    open societies, 154
Bellis, M., 96, 222
Ben & Jerry's, 258
Benner, P., 50
Bennett, C. A., 10
Bennett, I. J., 51
Benson, Nick, 188
Beppu, T., 61
Beyeler, W., 61
Bhardwaj, M., 65
Binderup, K., 171
Bio-based industrial processes, 78
Bioethical issues, 65
Bioethics, 65
    National Bioethics Advisory Commission, 69
Bioleaching, 76
Biometric measures, 94
Biopharmaceuticals, 70
    advantages of, 72–73
    challenges related to, 74–75
    drugs for rare diseases, 75
    emergence of, 71
    genetic engineering and, 72
    hybridoma technology, 72
    patents, 73–74

*Index*

physical performance enhancements, 75
Biotechnology
  agricultural, 65–70
  animal cloning, 69
  consumer's right to know, 67
  defined, 60
  early history of, 61–63
  energy and power, 81–87
  environmental, 75–77
  ethical issues and, 65–70
  Genetically Engineered Food Alert, 67
  historical timeline of, 62–63
  industrial, 77–81
  information and communication, 87–96
  instructional activities, 193–196
  modern era of, 64
  pharmaceutical, 70–75
  science and engineering in relation to, 193
Bird, S., 40
Blake, R. H., 52
Blanchard, Ken, 255
Blanchard, S. M., 51
Borgmann, A., 151
Borlaug, N., 66
Bottorff, W. W., 96
Brandt, R. B., 127
Braungart, M., 136
Breeder reacting technology, 81
  ethical considerations, 83
  history of, 82–83
Bronowski, J., 155, 156
Bronzino J. D., 51, 53
Brooks, R. R., 76
Brown, Danny C., 145
Brusic, S. A., 167
Buchholtz, A. K., 253, 259, 261
Budgeting and ethics, 105
Bunch, L. G., 25
*Business Ethics*, 258

Business for Social Responsibility, 257
Byron, S. J., 22

**C**

Callahan, D., 56
Calvinistic beliefs, 25
Campbell, K., 128
Capitalism
  Protestant work ethic and, 26
Care-based thinking, 5
Career ethic, 28
Carroll, Archie B., 243, 245, 249, 251, 253, 259, 261
Case studies, 50–113
  *see also* Instructional activities
Casey, L. B., 55
Caste system, 37
Categorical imperative, 15
Center for Ethics in Government, 260
Character
  developmental and contextual issues related to, 145
Chelliah, D., 152
Cherry, M., 28
Chiger, S., 95
Citizens
  technologically literate, 211
Clark, K. B., 230
Clean Air Act, 100
Codes of professional conduct (ethics), 53, 228
Communication and information technology
  *see* Information and communication technology
Computer Ethics Institute, 248
Construction technologies, 107–113
  and ethics, 108
  community planning, 109
  environmental concerns, 109
  ethical issues related to, 109
  instructional activities, 235–240
  Pleasant City Jail case, 110–112

*Index*

public services, 109
socioeconomic concerns, 109
Consumption
  economics and ethics of, 153
  responsible, 153–154
Consumption and technology, 151
Conway, R., 171
Corporate citizenship, 258
Corporate ethics, 257–260
Corruption, 33
Council for Secular Humanism, 28
Council on Technology Teacher Education (CTTE), 179
Craft ethic, 28
Cruse, J. M., 52
CTTE
  *see* Council on Technology Teacher Education
Cua, A. S., 36
Cultural beliefs, 22
Cultural genealogies, 145
Culturally diverse technological world
  ethics in, 21
Cultural value systems, 146
  and change, 147
Culture, 2
  defined, 22
  dominant, 31
  effects of technological progress on, 123
  eroded by technology, 150
  ethnic, 30
  relationship between ethics, values, morals, and, 23
Curran, W. J., 50
Custer, Rodney L., 145
Cyberterrorism, 95
Cynicism, 126
  technology and, 126

**D**

Dautheribes, J. L., 13
Dean, H. R., 167
Death

case of Karen Ann Quinlan, 55–56
definition of, 54
Debate
  ethics, 133
  ethics and values, 213–215
  evaluation, 138, 140
  procedures, 139
  rubric for evaluating, 141
  topics, 137–138
DeCarvalho, R. J., 147
Deculturation, 31
DeMiranda, Michael, 50, 60, 188, 193, 227
Democratic principles
  and technology-related issues, 157–158
Deontological perspectives, 4
Deontologist, 4
Deontology, 4, 273
Department of Homeland Security (DHS), 93
Descriptive relativism, 127
Determinists, 129
Devier, D. H., 178
DeVore, P. W., 163
Dewey, G., 167, 172
Dewey, John, 7, 158
DHS
  *see* Department of Homeland Security
Dickerson, S. S., 54
Dinan, Stephen, 94
Distribution of wealth, 134
DNA detector, 57
  detection of biological warfare agents, 57
Drenth, P., 42
DuBois, J. L., 153
Dugger, W. E., 166
Dyck, A. J., 50

**E**

Eckersley, R., 152
Economics

*Index*

and ethics of consumption, 153
Economic system, global
    ethics and, 243–264
Ellis, George F. R., 43
Empathy
    as ideal in ethical thinking, 157
Enderle, G., 36
Enderle, J. D., 51
Energy and power technologies, 81–87
    arguments for and against, 85
    breeder reacting, 81
    instructional activities, 196–211
    nuclear power technology, 81
    sustainable growth and, 199
Energy efficiency
    calculating, 197
    calculating total system, 198
Energy Policy and Conservation Act, 100
Energy ratios, 196
    calculating energy efficiency, 197
    calculating time to double energy consumption, 200
    calculating total system efficiency, 198
Enlightenment, 25
Enron, 244
Entrepreneurial ethic, 28
Environmental biotechnology
    bio-based substitutes for petroleum products, 77
    bioleaching, 76
    case of sewage sludge disposal, 76
    ethics and, 75–77
    hyperaccumulators, 76
Environmental Protection Agency (EPA), 68, 76
EPA
    *see* Environmental Protection Agency
Ethical behavior, 2
    conscience as source of, 2
    guilt as source of, 2
    intuition as source of, 2
    reason as source of, 2
Ethical beliefs, 2
    and culture, 2
Ethical challenges, 43
Ethical decision, 2
    model in technology education, 14
Ethical decision making, 251–257
    applying the Kidder model, 271–274
Ethical employees
    importance of, 245
Ethical genealogies, 145
Ethical issues
    in a technological world, 1–16
    overconsumption as, 152
Ethical norms, 129
    hierarchicalism and, 130
Ethical principles, 1, 251
    sources for, 2
    universal, 8
Ethical relativism, 3, 8, 13, 126
Ethical standards, 159
Ethical tests, 253
    common sense, 253
    making something public, 253
    one's best self, 253
    purified idea, 254
    ventilation, 253
Ethical thinking
    empathy as ideal in, 157
    self-discipline as ideal in, 157
Ethical universals, 129
    hierarchicalism and, 130
Ethical worldviews
    and impact on technology, 125
    cynicism, 126
    hierarchicalism, 129–131
    materialism, 128–129
    relativism, 126–128
    skepticism, 125
Ethics
    accommodation, 36
    agricultural biotechnology and, 65–70
    as facet of technological literacy, 49

assessment of technological
   impacts on society, 123–142
attributes related to, 10
bioethics, 65
budgeting and, 105
computer, 248
construction technologies and,
   107–113
consumption, 153
corporate, 257–260
debate of values and, 213–215
defined, 1
developmental and contextual
   issues related to, 145
energy and power technologies
   and, 81–87
environmental biotechnology and,
   75–77
for citizenship in a technological
   world, 267–275
immigration and American, 29–31
in a culturally diverse technological
   world, 21–44
in Africa, 33–35
in a global economic system,
   243–264
in Asia, 35–37
incorporating in curriculum, 165
industrial biotechnology and, 77–81
in Europe, 37–39
in government, 260–262
in Latin America, 39–40
institutionalizing, 259
instruction, 134–142, 164
in technology education literature,
   170–173
international perspectives on, 31–40
international technology education
   and, 174
in textbooks, 163
in the workplace training seminars,
   244
in United States of America, 22–24
management, 245–251

manufacturing technologies and,
   101–107
materials in modular activities, 166
medical advances, technology, and
   51–52
medical technology and, 52–53
normative, 129
pharmaceutical biotechnology and,
   70–75
philosophy of, 2–5
pluralistic views of, 155
professional, 176
relationship between culture, values,
   morals, and, 23
role in technology education, 10
strategies in technology education
   to include, 174
study of designed world and,
   187–240
teaching, 7, 11
technological development and,
   40–44
technology education and, 163
tests, 251–257
transportation technologies and,
   96–101
virtue, 5, 36, 252
work, 9
workforce, 243–245
Ethics in Government Act of 1978, 261
Ethics instruction
   debates as technique for, 136–142
   extent of, 174
   strategies for, 12
Ethics principles, 251–257
   principle of justice, 251
   principle of rights, 251
   principle of utilitarianism, 251
Ethics-related curriculum, 165
Ethics-related outcomes
   measurement strategies for, 169
Europe
   business ethics in, 37
   corporate value system, 39

## Index

ethics in, 37–39
Evans, W. E., 70
Expert systems, 151
Extreme materialism, 128

### F

Feminist theory, 252
Fiechter, A., 61
Fleischman, A., 59
Ford, S., 77
Foreign Corrupt Practices Act, 261
Founding Fathers
    Adams, John, 26
    beliefs of, 24–26
    Jefferson, Thomas, 26
    Madison, James, 26
    Washington, George, 25
Freas, M., 59
FreedomCAR program, 98
Freeman, S., 222
Frigitemp Corporation, 246
Fuller, R. B., 155
Furnham, A., 27, 28

### G

Geisler, N. L., 129
General Mills, 258
Genetically Engineered Food Alert, 67
Genetic engineering, 64
    animals, 66
    biopharmaceuticals and, 72
    crops, 66
    hybridoma technology, 72
Genetic profiling, 130
Gibbs, W. W., 44
Gilligan, C., 6
Glantz, L. H., 130
Glazner, P. L., 156
Global economic system
    ethics and, 243–264
    setting rules for, 261
    workforce ethics, 243–245
Global war on terrorism, 92

Gluckman, M., 34
Golden Rule, 252
Goldsborough, R., 248
Gomez, J. E. A., 39
Goodman, N., 36
Goree, K., 2
Gorman, M. E., 173, 227, 228, 229
Government
    ethics in, 260–262
Gradwell, J. B., 172
Grant, G., 159
Greenhouse effect problem, 99
Griffiths, M., 77, 78, 79
Gudkov, L., 33

### H

Hales, J. A., 11
Hands-on experiences in teaching, 187
Hanson, R. E., 178
Hasbro, 258
Hays, R. H., 230
Hierarchicalism, 129–131
    ethical norms, 130
    ethical universals, 130
Helwege, A., 56
Henak, R., 107
Hendricks, J. L., 172
Hernandez, T., 76
Heroic efforts, 50
HEVs
    *see* Hybrid electric vehicles (HEVs)
Heywood, V., 71
Hill, Roger B., 1, 10, 12, 167, 172, 180, 267
Hippocratic Oath, 53
Hoban, T. J., 67
Hoffman, B., 51
Hofstede, G., 37
Hollman, A., 51
Homeland Security Act (HSA), 93
Homogenization, 150, 154
Honda Motor Company, 247
Honeywell, 258

## Index

Hopkins, W. E., 21, 22, 23, 31, 36, 38
Howell, J. D., 51
HSA
   *see* Homeland Security Act
Hughes, Angela, 142, 163
Human adaptive systems model, 11
Human development model, 146–150
   cultural dimension, 147–148
   economic dimension of, 146–147
   *illus.*, 147
   institutional dimension, 148
   technology and, 149–150
Humber, J. M., 55
Hussain, M., 32
Huyke, G. J., 173
Hybrid electric vehicles (HEVs), 97–99
   FreedomCAR program, 98
   instructional activity, 221–227
   Partnership for a New Generation of Vehicles, 97
Hybridoma technology, 72
Hyperaccumulators, 76

## I

IACP
   *see* Industrial Arts Curriculum Project (IACP), 166
Immigration, 31
Immigration Act of 1990, 29
Immigration and American ethics, 29–31
Immoral management
   examples of, 246–247
Industrial Arts Curriculum Project (IACP), 166
Industrial biotechnology
   bio-based industrial processes, 78
   ethical challenges to, 79
   ethical issues in, 77
   impediments, 78
   structural resistance to, 78
Information and communication technologies, 87–96
   balance between freedom, privacy, and security, 91–96
   Department of Homeland Security (DHS), 93
   environmental costs of, 88
   global war on terrorism and, 92
   Homeland Security Act (HSA), 93
   personal communication systems pollution, 88–89
   right to privacy, 94
   spam, 89–91
   visual pollution, 88
Information-intensive technological developments, 145
Informed consent, 50
Inglehart, R., 146
Insider trading, 105
Instructional activities, 187–240
   a chance for a future—grade levels 11–12, 193–196
   a retrieval device, 231–235
   biotechnologies, 193–196
   ethics and values debate, 213–215
   ethics in a global economic system, 262–264
   hybrid electric vehicle research, 221–227
   information and communication technologies, 211–220
   medical technologies, 188–192
   promotional campaign, 215–219
   public opinion poll, 224–227
   sound in a box—grade levels 3–7, 189–191
   the energy game, 201–211
   the ethics are in the specs, 236–240
   the stethoscope—grade levels 8–9, 191–192
   transportation technologies, 221–227
   Webquest: ethics and technology in the classroom, 219–220
International perspectives on ethics, 31–40
   anti-American sentiments, 32

*Index*

International Society for Technology in Education (ISTE), 168
International Technology Education Association (ITEA), 11, 124, 134, 164, 187, 267
    Center for the Advancement of Teaching Technology Science, 173
ISTE
    *see* International Society for Technology in Education (ISTE)
ITEA
    *see* Internal Technology Education Association (ITEA)

## J

*Jackson's Mill Industrial Arts Curriculum Theory*, 11
Järvinen, E-M., 175
Jefferson, Thomas, 26
Jencks, C., 151
Johnson, D. G., 3, 4
Johnson & Johnson, 259
Josephson Institute of Ethics, 9, 157, 179, 244
    ethics in the workplace training seminars, 244
Juma, C., 65
Justice
    compensatory, 251
    distributive, 251
    procedural, 251
Justice theory, 15, 273

## K

Kant's categorical imperative, 15
Karsnitz, J. R., 172
Kate, K. 77, 78
Kendall, P. A., 67
*Kenosis*, 43
Kidder, R. M., 5, 8
    common values, 8
ethical decision-making model in technology education, 13, 14, 271
Kinnier, R. T., 13
Klingemann, H., 146
Kohl, M., 55
Kohlberg, L., 6, 7, 12, 16
Kohn, A., 156, 157
Kotzar, G., 59
Kreeft, P., 126
Kristiansen, B., 64
Kryl, D., 79
Kurtz, P., 28

## L

Laird, S. A., 77, 78
Lalan, S., 52
LaPorte, J. E., 167
Latin America
    ethics in, 39–40
    nepotism, 39
    value system in, 39
Lee, C., 95
Lemos, R., 95
Levinovitz, A. W., 51
Levi Strauss, 258
Lewis, C. S., 9
Lewis, S., 100
Life and death decisions, 53–56
Liker, J. K., 227
Litowitz, Len S., 81, 196
Living wills, 50
Lotus Development Company, 258

## M

Maccoby, M., 28
Macer, D. R. J., 64, 65, 68
Madison, James, 26
Magesa, L., 34, 35
Management
    amoral, 249–250
    archetypes, 245
    immoral, 245–246

individual hypothesis, 250
moral, 247–248
patterns of, 250–251
population hypothesis, 250
Management ethics
approaches to, 245–251
Manufacturing education, 228–230
Manufacturing educators, 230
Manufacturing Enterprise class, 105–106
Manufacturing technologies, 101–107
case of defective cords, 103–105
corporate greed, 106
ethical issues, 102, 103
instructional activities, 227–235
internal corruption, 106
making a fair profit, 105–106
offshore factory operations, 102
poor work ethic, 106
safety of fuel tanks, 102
Marcus, A. I., 153
Markert, Linda Rae, 21, 42
Materialism, 128–129
extreme, 128
impacts on technology, 129
Matsumura, M., 28
Mauzur, D. J., 52
May, W. T., 157
McCullough Corporation, 248
McDonald's, 250
McDonough, W., 136
McElroy, J. H., 22, 23, 24
McLaughlin, R., 39
Meche, M., 7
Medical advances, technology, and ethics, 51–52
Medical technologies
acts of mercy, 50
codes of professional conduct and, 53
ethical dilemmas and, 50
ethics and, 52–53
future of, 56–57

genetic engineering, 64
heroic efforts, 50
informed consent, 50
instructional activities, 188–192
life and death decisions and, 53–56
medical advances and ethics, 51–52
micro electromechanical systems (MEMS), 58–60
nanomedical case, 57–58
preventative, 56
quality of life, 50
recombinant DNA (rDNA), 64
rescue, 56
MEMS
see Micro electromechanical systems
Merck, 248
Metaethical relativism, 127
Meyyappan, M., 57
Micro electromechanical systems (MEMS), 58–60
Mill's principle of utility, 15
Mitchell, E., 96
Modular technology education
ethics materials in, 166
Moral development theory, 5–6
Kohlberg's stages of, 6
Morality, 23
pluralistic views of, 155
Moral law, 9
Moral management, 247–248
examples of, 248
Moral principles, 9
distinguished from personal opinions, 9
Moral relativism, 126
Morals
relationship between ethics, culture, values, and, 23
Morrison, R. S., 52
Mulgan, G., 228
Muller, H. J., 158

*Index*

## N

NACFAM
 see National Coalition for Advanced Manufacturing (NACFAM)
Nanomedical technology, 57–58
 DNA detector, 57
Nash, Laura, 255
Nash, R., 21
National Academy of Engineering and National Research Council, 172
National Bioethics Advisory Commission, 69
National Coalition for Advanced Manufacturing (NACFAM), 168
National Council for Accreditation of Teacher Education (NCATE), 177
National Council for the Social Studies, 170, 212
National Council of Teachers of Mathematics, 168
National Education Association (NEA), 178
National Science Teachers Association (NSTA), 168
*National Standards for Social Studies Teachers*, 170
Naturalism, 28
NCATE
 see National Council for Accreditation of Teacher Education (NCATE)
NEA
 see National Education Association
Negative rights, 4
Nice, K., 97
Nike, 258
Nisbett, R., 148
Nish, S., 9
Norman, D. A., 123
Normative ethic, 129
Normative relativism, 127

NSTA
 see National Science Teachers Association
Nykänen, J., 175

## O

Oates, W., 27
Olson, D. W., 166
Omnibus Communication and the Digital Millennium Copyright Act, 87
Organization for Economic Cooperation and Development (OECD), 40, 261
Orphans, 75
Ott, B. B., 55
Overconsumption, 152
Overington, J. P., 75

## P

Padover, S. K., 24
Paine, Lynn Sharp, 248
Palermo, G., 108
Pannabecker, J. R., 171
Parker, J., 222
Partnership for a New Generation of Vehicles, 97
Peale, Norman Vincent, 255
Penticuff, J. H., 54
Personal opinions
 distinguished from moral principles, 9
Petrina, S., 167, 171
Petty, G. C., 10
Pharmaceutical biotechnology, 70–75
Piaget, Jean, 7
Poe, J. B., 227
Pomerantseva, I., 52
Pond, R. J., 228
Positive rights, 4
Postman, Neil, 132
Postmodern approach to science, technology, and socicty, 157

Potsky, A., 169
Power
  managing, 154
*Power of Ethical Management (The)*, 255
Presley, Susan, 163
Preventative medical technologies, 56
  micro electromechanical systems and, 58–60
Principle of caring, 252
Principle of justice, 251
  compensatory justice, 251
  distributive justice, 251
  procedural justice, 251
Principle of rights, 251
Principle of utilitarianism, 251
Principles of Medical Ethics, 53
Professional ethics
  in technology education, 176
Protestantism, 25
Protestant work ethic
  *see* Work ethic
Puritan ethic, 28

## Q

Quality of life, 50, 152–153
Quinlan, Karen Ann, 55–56

## R

Rajput V., 52
Ratledge, C., 64
Rawl's theory of justice, 15
rDNA
  *see* Recombinant DNA technology
Recombinant DNA (rDNA) technology, 64, 72
Reebok, 258
Reed, Philip A., 163, 166, 171
Regan, T., 155
Reid, M., 175
Reiser, S. J., 50
Relativism, 3, 8, 126–128
  descriptive, 127
  metaethical, 127
  moral or ethical, 126
  normative, 127
Relling, M. V., 70
Rescue medical technologies, 56
*Revolution in Progress: Today's Proliferating Science and Technology*, 151
Rifkin, J., 64
Rights
  human, 5
  legal, 5
  negative, 4
  positive, 4
  principle of, 251
Right-versus-right dilemma, 15, 272
Ringertz, N., 51
Roberts, P., 95
Roche, P. A., 130
Rokeach, M., 22
Rosenblatt, J., 53
Rubric
  for scoring classroom debates, 86, 141
  for scoring multimedia project, 194
Russell, R. S., 230

## S

Sanders, Mark, 87, 167, 171, 211
Sarkees-Wircenski, M., 11
SCANS
  *see* Secretary's Commission on Achieving Necessary Skills (SCANS)
Scott, J. L., 11
Sears, Roebuck and Company
  guidelines for making ethical decisions, 256
Secretary's Commission on Achieving Necessary Skills (SCANS), 168
Secular humanism, 28
Segal, H. P., 153
Self-discipline
  as ideal in ethical thinking, 157
Senge, P., 78

*Index*

Separation of church and state, 25
Seymour, Richard D., 101, 227
Shell Oil, 258
Shirouzu, N., 98
Shrader-Frechette, Kristin, 123
Sierra Club, 99
Singh, J. P., 37
Skepticism, 125
   technology and, 125
   tolerance of all viewpoints, 125
Smith, V. H., 51
Snellen, H. A., 51
Snyder, J. F., 11
Sobek, D. K., 227, 230
Social institutions, 146
Society
   closed, 154
   ethics and assessment of technological impacts on, 123–142
   impact of wireless communication devices on, 124
   open, 154
Spam, 89
Spammer, 90
   pornography, 90
Sport utility vehicles (SUVs), 99–101
   Clean Air Act and, 100
   Energy Policy and Conservation Act and, 100
   instructional activity, 224–227
   standardized bumper heights, 100
*Standards for Technological Literacy: Content for the Study of Technology,* 11, 49, 124, 134–136, 164, 168, 170, 187, 267
   ethics addressed in, 170
   ethics instruction and, 134–142, 228–229
Stephens, Diane Irwin, 163
Stout, D. A., 7
Stride Rite, 258
Sullivan, B., 94
Sullivan, P., 108

SUVs
   *see* Sport utility vehicles (SUVs)
Swartz, M., 244
Swearengen, J. C., 43
Swenson, W., 37
Swindells, M. B., 75
Synge, H., 71

**T**

*Tao,* 36
Taylor, B. W., 230
Teacher education
   aspects of ethics to be included in, 179
Teaching ethics
   contextual learning, 102
   debates as technique for, 84, 86, 136–142
   instructional activities, 136–142, 187–240
   key factors in, 11
   role of technology teacher in, 131–133
Technocrat, 132
Technological advances
   effect on civilization, 123
Technological change
   pace of, 151
Technological development
   ethics and, 40–44
   information-intensive, 145
Technological impacts on society
   ethics and, 123–142
Technological literacy
   areas in standards, 164
   as goal, 243
   eroding, 149
   ethics as facet of, 49
   help in advancing, 163
Technologically advanced culture, 152
Technologically literate individual, 172, 211
Technologically literate society, 60

Technological world
  ethics in a culturally diverse, 21
  role of technology teacher in, 156
Technology
  characteristics of, 51
  consumption and, 151
  cynicism's view of, 126
  discouraging citizen participation, 149
  eroding culture, 150
  hierarchicalism, 129–131
  human development and, 149
  interaction with culture and values, 146
  materialism, 128
  relativism's view of, 127
  role in balancing power, 158
  skepticism's view of, 125
  sustainable forms of, 124
Technology administrators
  ethical behavior by, 178
Technology education
  applying the Kidder model for ethical decision making, 271
  balance in, 173
  classroom debates in, 84, 86, 136–142
  defined, 11
  ethical decision-making model in, 14
  including ethics in, 174
  instructional activities, 187–240
  international, 174–176
  modular, 166
  professional ethics in, 176
  role of ethics in, 10, 174
  rubric for scoring classroom debates, 86, 141
  rubric for scoring multimedia project, 194
  status of ethics in, 163–182
  strategies for including ethics in, 174–176
Technology education literature
  ethics in, 170–173
Technology educator
  addressing questions of ethics and values, 212
  ethical behavior by, 177, 178
  role in empowering citizens in a technological world, 156
  role in teaching ethics, 131–133
*Technopoly: The Surrender of Culture to Technology*, 132
Teleology, 23
Terrorism
  airport security, 92
  cyberterrorism, 95
  global war on, 92
  Homeland Security Act (HSA), 93
  racial profiling, 92
Terrorist attacks, 32
Terzi, R., 28
Therese, M., 13
Thibodeau, P., 95
Timberland, 258
Todd, R. D., 172
Toffler, B. L., 244
Transparency International, 261
Transportation technologies, 96–101
  history of, 96–97
  hybrid electric vehicles (HEVs), 97
  instructional activities, 221–227
  sport utility vehicles (SUVs), 97
Tucker, E. M., 7
Turner, V. W., 34
Tyco, 244

## U

Unethical behaviors, 42
United States Department of Transportation (USDOT), 100
United States of America
  beliefs of Founding Fathers, 24–26
  ethics in, 22

*Index*

United States Office of Government Ethics, 261
United States Patent and Trademark Office (USPTO), 73
  biopharmaceutical issued, 74
Universal values, 8–10
Utilitarianism, 3–4, 273
  principle of, 251
Utility principle, 15

**V**

Vacanti, J. P., 52
Values, 22, 175
  defined, 1
  relationship between ethics, culture, morals, and, 23
  universal, 8–10
Vanderburg, W. H., 123
Vincent, A., 7
Virtue
  defined, 5
Virtue ethics, 5, 36, 252, 273
Visual pollution, 88
Vogel, D., 244, 259
Volk, K., 174

**W**

Wade, M. L., 51
Wagner, P. A., 178
Walsh, G., 71
Ward, A. C., 227
Warner, W. E., 166
Washington, George, 25
Wassman, B., 108
Watkins, S., 244
Weber, Max, 10, 26
Wells, J. G., 173
Wells, P., 100
Welzel, C., 146, 148, 150, 159
Wescott, Jack W., 107, 235
Westra, Laura, 123
Wheelwright, S. C., 230
Wicklein, Robert, 123, 131

Wiens, A. E., 177
Williams, D., 57
Williams, P. J., 175
Williams, Stephanie, 96, 221
Wireless communication devices
  impact on society, 124
Wolfe, A., 23
Womble, Myra N., 12, 96, 221
Woodhouse, E. J., 41, 43
Work ethic, 9, 26–28
  and capitalism, 26
  poor, 106
Workforce ethics, 243–245
WorldCom, 244
World Medical Association
  Hippocratic Oath, 53
  Principles of Medical Ethics, 53
Wright, J. R., 178
Wright, R. T., 166

**X**

Xenotransplantation, 66

**Y**

Yi, S., 174
Yung, J. E., 166

**Z**

Zajtchuk, R., 57
Zechendorf, B., 77
Zimmerman, F., 231